IET CONTROL, ROBOTICS AND SENSORS SERIES 115

Ground Penetrating Radar

The IET International Book Series on Sensors

IET International Book Series on Sensors–Call for Authors

The use of sensors has increased dramatically in all industries. They are fundamental in a wide range of applications from communication to monitoring, remote operation, process control, precision and safety, and robotics and automation. These developments have brought new challenges such as demands for robustness and reliability in networks, security in the communications interface, and close management of energy consumption. This book series covers the research and applications of sensor technologies in the fields of ICTs, security, tracking, detection, monitoring, control and automation, robotics, machine learning, smart technologies, production and manufacturing, photonics, environment, energy, and transport.

Proposals for coherently integrated international multiauthored edited or coauthored handbooks and research monographs will be considered for this book series. Each proposal will be reviewed by the IET Book Series Editorial Board members with additional external reviews from independent reviewers. Please e-mail your book proposal to: vmoliere@theiet.org or author_support@theiet.org.

Ground Penetrating Radar

Improving sensing and imaging through numerical modeling

X. Lucas Travassos, Mario F. Pantoja and Nathan Ida

The Institution of Engineering and Technology

Published by The Institution of Engineering and Technology, London, United Kingdom

The Institution of Engineering and Technology is registered as a Charity in England & Wales (no. 211014) and Scotland (no. SC038698).

First published 2021

The Institution of Engineering and Technology
Michael Faraday House
Six Hills Way, Stevenage
Herts, SG1 2AY, United Kingdom

www.theiet.org

British Library Cataloguing in Publication Data
A catalogue record for this product is available from the British Library

ISBN 978-1-78561-493-4 (hardback)
ISBN 978-1-78561-494-1 (PDF)

Typeset in India by MPS Limited

Contents

About the authors

X. Lucas Travassos is an assistant professor in the Joinville Technological Center at the Federal University of Santa Catarina, Brazil. His research interests cover the design and optimization of electromagnetic devices, electromagnetic compatibility and antennas and propagation.

Mario Fernández Pantoja is a full professor in the Department of Electromagnetism and Physics of Matter at the University of Granada, Spain. His research interests include the areas of time-domain analysis of electromagnetic radiation and scattering problems, radar technology, optimization methods applied to electromagnetics, Terahertz technology and nanoelectromagnetics.

Nathan Ida is a distinguished professor of electrical and computer engineering at The University of Akron, Ohio, USA. His research interests are in the areas of numerical modeling of electromagnetic fields, electromagnetic wave propagation, nondestructive testing of materials at low and microwave frequencies and in sensors and actuation with an emphasis on interfacing and integration.

Preface

The idea of detecting and identifying buried objects and materials (solid or liquid) is almost as old as human civilization itself. Among the many ingenious solutions designed for this purpose, the use of radar pulses has been the subject of intense research and development activities over the last 80 years or so, with significant progress. Thanks to advances in electronics, the development of portable systems led to a proliferation of commercial systems capable of identifying archaeological or geological structures, without the need of costly and invasive earthworks. In efforts to improve performance, some specific CAD software has also been developed to process the massive amounts of data generated from the dispersion and reflection of electromagnetic waves from various buried targets.

The initial applications envisioned for these early radars, dubbed ground penetrating radars (GPRs), were in geophysics and archaeology but it became apparent quite early on that other applications, notably in surveys of infrastructures such as roadways, pipelines, and other utilities were within the capabilities of even the simplest GPRs. Even more critically, GPRs were pressed into service in demining operations where performance is paramount. More recently, the need for nondestructive testing (NDT) and evaluation of critical structures, primarily made of reinforced concrete, has seen the adaptation of GPR to this task. This also entailed a different approach whereby the depth of penetration was less important than resolution, since now the search was for small targets or for corrosion products on rebars.

Although GPRs have become relatively common and have improved considerably in the last 30 years or so to the point that in some specific applications their performance is on par with other nondestructive methods, the fact remains that in many application GPRs perform poorly. One of the reasons for this state of affairs stems from the simple fact that successful predictions require highly qualified and coordinated multidisciplinary teams. Only from the confluence of information from physicists, archaeologists, geologists and electrical and mechanical engineers (and often others), is it possible to interpret the datagrams resulting from long hours of hard work in the field and provide answers to the simple questions of where and what is buried in the ground. Regrettably, as it has been often demonstrated in recent years, even high-performance GPR systems operated by inexperienced personnel lead to erroneous and costly predictions. What is more, sometimes those failures are often widely disseminated in cases with high social expectations and demands, as for instance in forensic applications. The conclusion is that GPR systems are technologically mature at the hardware level (of course many innovations are still possible and needed), but automated prediction algorithms are still far from being considered reliable.

The present work evolved from the authors' interests in a number of research areas, including antenna theory and design, radar applications, nondestructive evaluation (NDE), and numerical modeling and computation. This confluence of interests led to the idea of bringing the power of numerical models to bear on those issues in GPR that were felt to be lacking. Specifically, the need for accurate interpretation of targets and host media in terms of geometric features and evaluation of material properties with limited experimental data suggested that the missing data can be generated through a numerical experimental test bed. If one can generate signals from any material and target distribution, then surely one can match a specific distribution with data obtained from field surveys. That is the basic contribution of numerical models in the context of GPR. To make this process practical, additional tools are needed, including ways of selecting the appropriate solutions, optimizing the solution space and assessing errors. In this process, the numerical models become an element of the overall process of identifying, analyzing and imaging targets.

This process, which incorporates numerical modeling tools in the process, is not without cost. It increases computation and requires additional algorithms and the development of these algorithms. However, this is done on the software level and separate from the data acquisition process. As such it has an added value that fully justifies the effort. This effort can also increase the reliability of detection, can improve imaging and in the end, the confidence in GPR itself.

The text in front of you contains six chapters in which we discuss fundamental issues from engineering, physics and computation using our experience in teaching and research in the field, trying to avoid vague statements and generalizations. After introducing the main concepts associated with GPR in the first chapter, we discuss the electromagnetic theory necessary to understand the problem complexity and as a preparation to addressing antennas and elements of antenna theory in Chapters 2 and 3. Chapters 4 and 5 are devoted to the system specification and numerical modeling, respectively, and finally in Chapter 6 we discuss pattern recognition techniques with the purpose of improving imaging.

Theoretical and mathematical background is contained in Chapters 2 and 5. Elementary definitions and properties of time-dependent and time-harmonic wave equations and their behavior in various media are discussed on the basis of Maxwell equations. This consists essentially, from the theory point of view, in finding the solution for given configurations, taking into account important system characteristics such as the frequency and the properties of the GPR assessment environment and the changes that can take place in various applications. For a given real-world GPR problem, proper material characterization poses critical constraints on evaluation since material properties impact the attenuation and phase constants as well as dispersion. No proof of the solution of Maxwell equations for the GPR problem with all its complexity is given in the present book relying on the fact that the reader can find more details on the solutions for complex media in the extensive list of references provided. However, the content of these two chapters is sufficient to render the text self-contained with regard to important aspects related to numerical modeling.

If there is any single component of radar that is obvious, it must be the antenna. All interaction with the test environment is through the antenna. The importance of the antenna to the problem is discussed in Chapter 3. After discussing the properties of elementary antennas, which is an essential step toward any type of antenna design, the most common GPR antenna designs in use are discussed and methods to improve their performance by optimization via stochastic algorithms are presented. The essential tool used in this step is a genetic algorithm capable of producing a Pareto front with multiple solutions for a multi-objective conflicting problem. Algorithms of this kind are constructed using multiple dimension problems that could not be simply visualized or understood. Various examples are given in order to illustrate the theory and antenna diversity; however, the practical situations arising in the implementation of those devices are not discussed. By limiting the discussion to the essential details, the text intends to present a clear picture of what is needed for the numerical representation of the core device in a GPR system. A reader interested in GPR will find in this chapter a useful introduction to the topic. For antenna designers, the chapter can also be useful to point out important aspects that are specific or important in GPR applications.

The major emphasis of Chapter 4 is related to GPR system specification. The approach is to take the system apart and analyze its individual blocks. This is essential for effective understanding of the system as a whole but also allows discussion of specific issues that are most important for GPR surveys or are important in assembling GPR systems. The major technical advice relates to a discussion of discrete-time systems and the impact on the data collected from surveys. It is quite technical but does include introductory material on signal and systems. The chapter concludes with a discussion on signal processing algorithms and their application to real-world data.

After discussing the general issues of GPR, the book embarks on a detailed discussion of numerical modeling. The discussion is intended to be inclusive so that by the end, the reader should be capable of identifying and applying the technique that best suits his or her goals. As pointed out in Chapter 2, there are several constraints and assumptions regarding the solution of Maxwell's equations for real-world problems. But the development of formulations and numerical methods, along with fast computers brought fast computation times for complex problems. This led to the techniques discussed in Chapter 5. The goal is a better understanding of the variables in the numerical representation process that cannot always be achieved in commercial software. Although numerical techniques are well developed, they are still progressing. In this context, Chapter 5 presents an innovative approach to speed GPR simulations by optimizing the boundary conditions. In this sense, one can view numerical techniques and its resulting optimal algorithms as a branch of a multidisciplinary problem solving approach which to a large extent is the subject matter of this book.

Pattern recognition has developed extensively over the past 20 years. In Chapter 6, emphasis has largely been in applying different techniques considering their limitations and drawbacks for GPR applications. Although pattern recognition theory has developed apart from the field of GPR, their combination appears to be a perfect match for eliminating the necessity of human operators to interpret the data. In retrospect, one can view pattern recognition and its developments as a branch of signal processing. Chapter 6 presents some developments implemented over the past

15 years by the authors to reduce the complexity of interpreting large amounts of data using different approaches ranging from simple optimization algorithms to neural networks and feature selection.

The topics covered in this text and the extent of coverage of the topics represent the complexity of the problem. It is hoped that the coverage will be helpful to practitioners and researchers in this extraordinarily important field.

Acknowledgments

Finally, we would like to gratefully acknowledge the contributions to this work by Dr. Douglas Viera (ENACOM—Handcrafted Technologies, Brazil), Prof. Sérgio Ávila (Instituto Federal de Santa Catarina), Prof. Paulo Aranha (Universidade Federal de Minas Gerais), Prof. Leone Peter Correia Andrade (Centro Integrado de Manufatura e Tecnologia), Prof. Rafael Gomez-Martin and Prof. Amelia Rubio-Bretones (University of Granada, Spain), Prof. Alexander Yarovoy (Technical University of Delft, Netherlands), Prof. Rodney Rezende-Saldanha (Universidade Federal de Minas Gerais, Brazil), and Dr. Peter Meincke (Denmark Technical University, Denmark).

Xisto Lucas Travassos, Joinville, SC, Brazil
Mario Fernandez Pantoja, Granada, Spain
Nathan Ida, Akron, OH, USA
September 2020

Chapter 1
Introduction to ground penetrating radar

1.1 Introduction

Ground penetrating radar (GPR) has been around as an idea for almost 100 years and as a practical method of probing inside media for well over 40 years with various degrees of success in disciplines ranging from prospecting to archaeology. In some ways, it may be considered a mature technology, at least as far as the hardware involved. Whereas it is usually associated with probing of soils for buried target, it also serves as a tool, uniquely adapted to nondestructive testing (NDT) and evaluation (NDE) of structures, especially those made of concrete (nuclear power plant containment structures, roadbeds, airport runways), and where surveys over large areas must be carried out such as in utility work (pipelines, service tunnels). In some of these applications, GPR replaces existing methods such as ultrasound because of its ability to scan areas much faster. But there are other reasons for the success of GPR in testing in comparison with other established NDT methods; the equipment needed is highly portable, often self-contained, comes in a variety of configurations from small hand-held devices to vehicle installed equipment, it can scan at speeds compatible with driving speeds (e.g., in testing roadways). In addition, the equipment can be adapted to a variety of applications, power consumption is modest, including battery-operated options, and all that at a reasonable budget.

Nevertheless, ground penetrating radar, its construction, its operation, and most of all interpretation of results are not trivial issues. Selection of appropriate hardware, (including antennas, frequency of operation, and waveforms) can only be done by proper analysis of the specific application or class of applications to be undertaken. This requires evaluation of the host media involved (soils, concrete, etc.) and the targets one is trying to detect and image. Information on the type of targets, relative sizes one expects to detect, depth at which they are expected to be found, their composition and electric properties, expected contrast with the host medium and the like, all affect the choice of equipment, methods of data acquisition, and any subsequent data processing that may be needed. In addition, proper preparation for a survey should take into account sources of noise (internal to equipment and from external sources) clutter, reflections from nontarget sources, as these affect the signal-to-noise ratio (SNR), and hence the detection and evaluation processes. Many of these aspects of the survey may only be estimated but others may be accurately known.

For example, properties of some soils, concrete, specific types of rocks, and the like, have been measured and are well documented. The more that is known about a survey, the better the outcome. Clearly then, any GPR survey must be considered a multidisciplinary undertaking that involves electrical engineers, physicists, chemical engineers, earth scientists, computer scientists, and other engineers, all in an overall attempt to provide proper predictions from information collected during the survey.

This chapter introduces the subject as a multidisciplinary process that includes hardware and software components. It starts with the electromagnetic (EM) phenomena on which the method is based and the equipment constructed. Because one is not expected to build his or her own equipment, some of the characteristics of available commercial equipment are listed to give an idea of the state of the art. Finally, the chapter introduces some of the more important techniques for detection and classification of targets. These subjects will be expanded on in subsequent chapters but the discussion here should form an overall view of the issues involved in GPR assessments.

The urge to see the unseen is as old as humanity but the specific idea of mapping buried objects (targets) in the soil (host medium) through the evaluation of scattered EM waves was first introduced by Hulsenbeck in 1926 [1]. Further developments, primarily in remote sensing, led to a consistent theoretical foundation of the problem, thereby revealing the considerable potential of GPR as a detection technique [2,3]. Before the method could be practical, many more requirements had to be addressed. The transmission of EM waves through the host medium to a reasonable depth has to be efficient with practical power levels at the transmitter, the contrast between targets and the host medium must be sufficient to generate clearly detectable scattered fields, and the bandwidth of the transmitted pulses must be wide enough to provide the required resolution in terms of target dimensions. These requirements are indeed challenging but they are entirely in line with the challenges of any practical NDT method, whereby one tries to detect a target of any size in any medium and any depth under conditions that do not necessarily favor detection (such as low contrast or high attenuation of signals). In the very nature of any method is a set of limitations against which the developer and the user operates.

GPR is widely used in subsurface surveys exactly because it can overcome some of these limitations. At present, it is considered one of the most powerful NDE tools in comparison with seismic, acoustic, infrared, and optical sensing systems. The advantages offered by GPR are related to the overall performance achieved by high-frequency EM waves (microwaves): with reasonable power requirements, they combine a reasonably large penetration range (larger than optical and infrared but lower than acoustic and seismic methods) with sufficient spatial resolution (greater than acoustic and seismic but lower than infrared and optical techniques) and a fair contrast between host and target media. This ability to detect a wide range of targets has its drawbacks. Reflections from inhomogeneous soils and scattering from nontarget structures (known as clutters) interact with the scattered field of actual targets, masking the pulses and hence interfering with detection of targets. There are of course methods of mitigating these problems, including scanning at various frequencies, the use of two or three-dimensional scans, as well as methods of processing the data. These lead to the need of collection of large amounts of data from multiple measurements, requiring specialized post-processing tools to complete the survey.

The aim of this chapter is to describe the most relevant GPR features, the EM wave propagation and scattering phenomena involved, and some of the latest techniques used to improve the mapping of buried objects in host media. The following section is an overview of the GPR scenario and provides background information relevant in the design of surveys. Section 1.3 is a very brief introduction to the theory of EM waves, the issue of material properties, antennas, and other characteristics of commercial GPRs, including system specification issues. An introduction to and some specific software support tools are given in Section 1.4. The goal of these tools is to improve the performance and success rate of GPR surveys and is discussed here as a means of emphasizing their importance. As appropriate for an introduction, none of these subjects is treated in any depth—that will come in subsequent chapters in this text.

1.2 Overview of a GPR system

A typical GPR scenario is shown in Figure 1.1. The GPR operation may be described in four broad steps:

1. EM waves from the transmitting antenna are directed into the host medium under test;
2. typical reflection–refraction phenomena occur at interfaces, whereby the scattered wave reflects part of the incident energy, and the remainder energy travels through interfaces at a different velocity to greater depths in the host;
3. the refracted waves propagate through the host medium reaching a boundary of different electrical characteristics (called flaw or target);
4. this scattered wave (also called echo) travels back from the target to the receiving antenna.

The physics of GPR is based on the reflection, refraction, and scattering processes that transform the incident wave into a received signal with an amplitude and phase different from the transmitted wave. The particular received waveform is determined

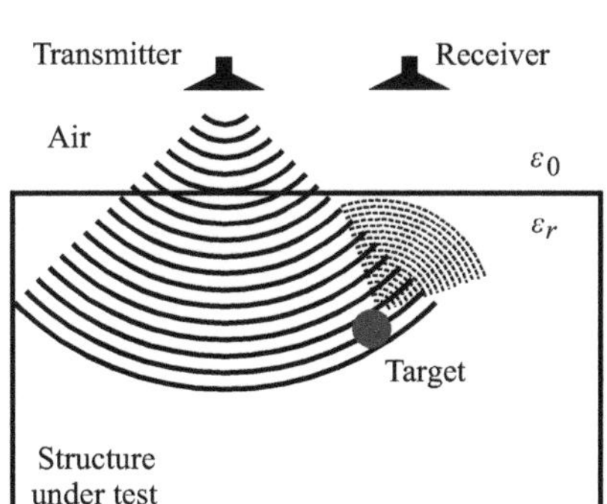

Figure 1.1 A generic GPR problem

primarily by the electrical contrast (i.e., ratios of relative permittivity and conductivity of targets and host media) as well as the particular dimensions and shape of targets.

Figure 1.2 shows a typical time-domain signal detected in GPR surveys. It can be described as the convolution of a number of waveform components, each representing the response of one component of the GPR system depicted in Figure 1.1, with the addition of noise, originating either from small flaws or inhomogeneous constituents of the host. The received waveform $f_r(t)$ can be decomposed as follows [4]:

$$f_r(t) = f_a(t) + f_s(t) + sg(t) + ns(t) \tag{1.1}$$

where $f_a(t)$ represents the effect of the antenna on the signal, $f_s(t)$ the reflections from the ground surface, $sg(t)$ is the signal from underground targets, and $ns(t)$ is the noise signal. In GPR measurements, $f_a(t) + f_s(t)$ are known together as the direct wave or clutter. The direct wave is viewed as undesired in the sense that it can mask signals from actual targets.

In GPR surveys, the process of identifying the scattering objects once a set of incident and scattered waves is known (in one or, most commonly, multiple locations in the host medium) is called the inverse problem, in which the received signals are processed for detection, classification, and localization of targets. To this end, advanced post-processing support tools (PSTs) such as neural networks [5], image processing [6], or wavelet transforms [7], as well as other well-known techniques such as filtering or amplification at some frequencies of interest [4] can be effectively used. Additionally, received waveforms can be processed by combining spatial, frequency and/or time-domain signals, with the latest being the most common choice. An important operation, common to all these techniques, is the removal of the direct wave. This is usually performed prior to the application of any post-processing operation. In spite of the many efforts over many years, accurate predictions in real-world scenarios are still wanting and improvements are still being sought. Some of the reasons for this unsatisfactory situation are that many of the more complex scenarios involve inhomogeneous or dispersive lossy dielectrics, as well as irregular boundaries

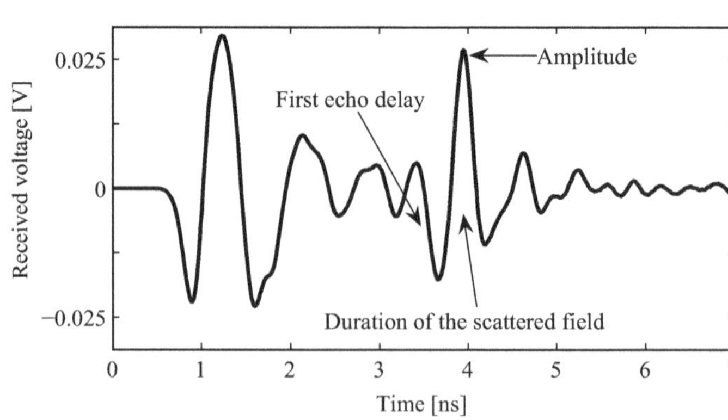

Figure 1.2 Reflected waveform from a buried target showing some characteristics of the waveform

in the host medium. Some of these difficulties can be alleviated through proper choice of equipment, testing strategies, and post-processing software.

Manufacturers of commercial GPR systems employ various approaches to assemble their systems, which can be tailored according to the application and target characteristics. In general terms, the main features to be considered in the variety of the GPR systems available are as follows:

1. unit size and weight;
2. power consumption;
3. central frequency;
4. transmitted pulse waveform (in terms of duration and modulation); and
5. number, type, and position of sensors (antennas).

The type and position of antennas and the size and shape of the structure under test determine the amount and quality of measurements. Depending on the number and position of transmitting/receiving antennas, there are three possible operating modes:

1. monostatic, whereby a transmitting/receiving antenna is located at the same place over the material under test;
2. bistatic, whereby transmitting and receiving antennas are at different positions; and
3. multistatic, whereby multiple transmitters and/or multiple receivers, usually designed to form antenna arrays, can be located at different positions;

Any specific choice of configuration will affect the amount of data and their quality, because a larger number of antennas at closer distances will result in strong coupling between antennas, either through the direct wave or the first reflection in the structure under test, effects that jeopardize the removal of clutter in the post-processing stage.

Additionally, the antenna type has considerable effect on the performance of the GPR and quality of surveys. Highly directive antennas explore narrow zones of the host at deeper distances but, in practice, they also increase the number of scans required. Furthermore, the type of antenna limits the operational bandwidth, which has to be chosen carefully. Higher bandwidths can be achieved through ultrawideband (UWB) antennas [8] that avoid the problem of late-time ringing, defined as a delayed EM radiation, which increases the noise in the data. On the other hand, the greater the bandwidth the higher sensitivity to external EM interferences (e.g., broadcast antennas operating in the same frequency range as the GPR system) and the lower energy per frequency available for the scan.

In this sense, the depth of penetration and resolution are also directly related to the pulse waveform and its frequency content. Time-domain or pulsed radars, commonly used in practice, are based on the radiation of a particular waveform with a high bandwidth, which eases the identification of targets and flaws by the presence of scattered waves, in a process called time-windowing. On the other hand, frequency-domain GPR systems use frequency modulation, whereby a carrier frequency scans bandwidths with a fixed step. Stepped-frequency continuous-wave (SFCW) systems are available in commercial GPRs, and their use is on the rise due to advantages over pulsed radars in terms of shaping the power spectral density and a higher mean power.

In summary, the choice of any GPR equipment has to be made according to the particular application (or class of applications) and characteristics of the target and host media. This choice also involves a trade-off between resolution and penetration depth, both key factors for the success in the detection and classification of targets. In the following section, the EM propagation nature of GPR is briefly discussed prior to comparing current commercial options available.

1.3 Fundamental theory of GPR

The operating principle of GPR is based on the propagation and scattering of EM waves in matter, offering thus an alternative to NDE methods based on energy–matter interactions of mechanical waves such as ultrasound. Therefore, instead of dealing with mechanical properties, GPR is based on the varying properties of EM constitutive parameters (permittivity ε, conductivity σ, and permeability μ) and their interaction with EM waves to locate, identify, and image embedded targets in host media.

1.3.1 Electromagnetic wave propagation

The nature of EM propagation and scattering is derived from Maxwell's equations relating the electric field intensity **E**, [V/m], the electric flux density **D**, [C/m^2], the magnetic field intensity **H**, [A/m], the magnetic flux density **B**, [Wb/m^2], and their sources, the electric current density **J**, [A/m^2] and the electric charge density ρ_e, [C/m^3]. Considering a general medium $\{\varepsilon, \sigma, \mu\}$, Maxwell field equations in differential form are as follows [9]:

Faraday's law:

$$\nabla \times \mathbf{E} = -\frac{\partial \mathbf{B}}{\partial t} \tag{1.2}$$

Ampère's law:

$$\nabla \times \mathbf{H} = \mathbf{J} + \frac{\partial \mathbf{D}}{\partial t} \tag{1.3}$$

Gauss's law for the electric field:

$$\nabla \cdot \mathbf{D} = \rho_e \tag{1.4}$$

Gauss's law for the magnetic field:

$$\nabla \cdot \mathbf{B} = 0 \tag{1.5}$$

where it should be noted that on the right-hand side of (1.2) and (1.3), the time-varying electric and magnetic flux densities also act as sources of each other. This key feature enables the propagation of EM waves through media with or without electric current or charges. Derived from these equations, a set of boundary conditions for the tangential and normal components of the fields has to be satisfied at interfaces between different media. The GPR problem can be viewed as a wave propagation problem but within the confinement of boundaries. In fact, reflection, refraction, and scattering of EM waves at boundaries provide essential means for measuring the interaction between EM waves and the structure under test, and target properties are measured through the received amplitude of EM fields at different frequencies.

EM wave propagation in a bounded space is defined by the nature of the electric properties of materials in the medium. In GPR surveys, materials can be considered as inhomogeneous, lossy and dispersive media, with constitutive parameters being local scalar quantities dependent on the frequency f of the wave ($\varepsilon = \varepsilon(\omega)$, $\sigma = \sigma(\omega)$, $\mu = \mu(\omega)\}$, where $\omega = 2\pi f$[rad/s]). The relation between field intensity vectors and flux density vectors can be written as $\mathbf{D} = \varepsilon\mathbf{E}$, $\mathbf{B} = \mu\mathbf{H}$, $\mathbf{J} = \sigma\mathbf{E}$. These are the constitutive relations that link the field equations to media, and, in general, must be considered as nonlinear. By incorporating these relationships in Maxwell's equations, a wave equation for the electric field intensity **E** or magnetic field intensity **H** can be obtained. For time-harmonic sources and in the time-domain, the wave equation for the electric field intensity in source-free, linear, isotropic, homogeneous media is

$$\nabla^2\mathbf{E} = \mu\frac{\partial}{\partial t}\left(\sigma\mathbf{E} + \varepsilon\frac{\partial\mathbf{E}}{\partial t}\right) \tag{1.6}$$

It is, however, more useful at this stage to convert this equation into the frequency-domain using the phasor transformation ($\partial/\partial t \rightarrow j\omega$), since many of the properties associated with EM waves are defined based on the frequency-domain representation of Maxwell's equations. Doing so converts the electric fields into phasors (each field quantity is now complex, described by an amplitude and a phase), resulting in the following:

$$\nabla^2\mathbf{E} = j\omega\mu(\sigma\mathbf{E} + j\omega\varepsilon\mathbf{E}) \tag{1.7}$$

For lossless nonmagnetic media, as is often assumed for the host medium in GPR applications, the EM wave propagates at a velocity v [9]:

$$v = \frac{c}{\sqrt{\varepsilon_r}} \tag{1.8}$$

where ε_r is the relative dielectric permittivity of the medium, and $c = 3 \cdot 10^8$ m/s is the speed of light. As will be further explained, the wavelength $\lambda = v/f$ plays an important role in determining the resolution capabilities of GPR.

In practice, most of the surveys contain lossy materials, which can have either conduction losses (i.e., media with non-negligible conductivity σ) or polarization losses (i.e., media with complex permittivity $\varepsilon_c = \varepsilon' - j\varepsilon''$). On the other hand, when the host medium is lossy, the wave propagates with an exponentially decaying amplitude. As a consequence, the depth range of the survey can be seriously affected, and a careful choice of frequency range and transmitted power has to be made in the GPR system specification. At this point, we characterize the attenuation in lossy media with the attenuation constant α:

$$\alpha = \omega\left[\frac{\mu\varepsilon'}{2}\left(\sqrt{1+\left(\frac{\varepsilon''}{\varepsilon'}\right)^2}-1\right)\right]^{1/2} \tag{1.9}$$

measured in nepers/meter [Np/m], and the velocity of propagation will be determined through the phase constant β,

$$\beta = \omega\left[\frac{\mu\varepsilon'}{2}\left(\sqrt{1+\left(\frac{\varepsilon''}{\varepsilon'}\right)^2}+1\right)\right]^{1/2} \tag{1.10}$$

as $v_p = \omega/\beta$, where v_p is the phase velocity of the propagating wave. The phase constant β defines the movement of the phase front and is related to the wavelength, defined as the distance between two wave fronts separated in time by one period of the (sinusoidal) wave. Thus, for moderately lossy materials, the detectable size of targets is limited by the spatial resolution $\lambda = 2\pi/\beta$ [m].

For lossy materials, the attenuation and the phase constants are frequency dependent, and that causes distortion of the pulse as it propagates through the host media. To operate in these dispersive media, GPR systems use various types of modulations to generate propagating waves in a range of frequencies. The velocity of this packet of waves is called group velocity. In order to detect a target in a lossy medium, a number of different waves are transmitted and they arrive at the target location at different times and with different phase angles. Even though this solution increases the depth range and the resolution, it also influences the waveform of the detected signal, making it more difficult to interpret the results.

The polarization of plane waves is another relevant parameter for the performance of GPR. Given that the field intensity of a uniform plane wave has a direction in space which may either be constant or time varying, the polarization corresponds to the curve traced by the electric field intensity vector in space as it propagates. The trace can depict a straight line, an ellipse, or a circle, corresponding to linear, elliptical, and circular polarization, respectively. In general, each antenna can transmit/receive EM waves with a certain polarization, and both should be oriented properly to maximize the detected signal. However, in practice, the material property can change the polarization of plane waves and this will result in some losses due to polarization mismatches.

The reflection and refraction of EM waves are the main phenomena involved in the detection of any target. In a GPR survey (Figure 1.1), an incident wave propagating from one medium (typically air) reaches a different medium (structure under test or host). A fraction of the wave is then transmitted (or refracted) into the host, while another fraction is reflected. The proportion of transmitted power depends on the angle of incidence of the wave on the surface of the host medium and the intrinsic impedances of air and the host medium. The intrinsic impedance, also called wave impedance, has the following form in lossy media:

$$\eta = \left[\frac{j\omega\mu}{\sigma + j\omega\varepsilon} \right]^{1/2} \tag{1.11}$$

Similar reflection/refraction occurs in the host medium when the incoming wave impinges on a flaw/target of constitutive parameters that are different than those of the host. In addition, the wave is attenuated due to the lossy properties of the host, with the attenuation constant given in (1.9). As a result, one should assess the electrical material properties of the surveyed media to predict the amount of power required to detect targets of a minimum size at a maximum depth.

1.3.2 Material properties

As can be seen from the previous section, a major consideration in any successful detection depends on the correct estimation of the electrical properties of the structure under test. Interactions between EM fields and matter in GPR must be considered from a macroscopic point of view. Nevertheless, the constitutive parameters of the media are derived from a microscopic point of view. Matter consists of atoms or molecules which in turn contain positive and negative charges. Charges that can travel through matter are called free charges, and those that cannot move freely (mainly because of a strong interaction with charges of opposite polarity) are called bound charges. In general, materials can be characterized by the predominant response of the charges when an EM field is applied to the structure. As the GPR technique is generally applied to lossy dielectric structures, the relevant physical responses are as follows:

- Polarization. In this response, the bound charges respond to the external electric field by displacing and thus creating small induced dipoles, which are polarized in the direction opposite that of the inducing electric field. The permittivity of the material is primarily a measure of its polarization.
- Conduction. Free charges move through the matter in response to the external EM field, a process in which the material absorbs power as a result of microscopic interactions. The result is dissipation (loss) of power of the EM wave due to conductivity.

It should be emphasized that electrical properties are related to

1. frequency;
2. position;
3. intensity; and
4. orientation of the driving EM field.

Considering (1), some dielectrics exhibit dipolar polarization or have molecules that are permanent dipoles. When an external low-frequency driving electric field is applied to this material, the charges in these molecules have sufficient time to change their direction. The dielectric permittivity is then frequency dependent and the material is called dispersive. The time needed for dipoles orientation is called the Debye

Table 1.1 Electrical properties of some materials (at 100 MHz)

Material	Relative permittivity	Conductivity [S/m]
Air	1	0
Asphalt	2–4	10^{-2}–10^{-1}
Clay	2–6	10^{-1}–10^{0}
Concrete dry	4–10	10^{-3}–10^{-2}
Epoxy	3–4.5	2×10^{-3}
Glass	4–7	10^{-10}
Granite	5	10^{-8}–10^{-6}
Limestone	7	10^{-8}–10^{-6}
Mica	6	10^{-15}
Polypropylene	2.5–2.6	10^{-13}
Polystyrene	2.56	10^{-14}
Polyethylene	2.26	10^{-13}–10^{-17}
Porcelain	6–8	10^{-10}–10^{-12}
PVC	3.39–4.52	10^{-13}

relaxation time and the wave propagation in this medium is nonlinear in frequency. Examples of this type of materials are those with water content, as for instance concrete. Regarding (2), when the medium's behavior varies when a uniform driving field is applied at different locations, this medium is called inhomogeneous. In this case, the constitutive parameters vary with position in the space of interaction. In open-field GPR surveys, inhomogeneous host properties can lead to changes in the maximum depth of analysis at different locations. In terms of intensity (3), for most applications the polarization of the medium responds linearly with the amplitude of the incident field. For this reason, we will consider only linear response to amplitude. In the same line, there are materials that are not equally polarized for different orientations of the driving field (4), but they rarely need to be considered in GPR applications. Table 1.1 illustrates the electrical properties of typical materials encountered in GPR work at a frequency of 100 MHz.

Following the guidelines of the previous section, one can clearly infer that EM waves suffer less attenuation in some materials such as air or asphalt than in others such as concrete or clay, due to their lower conductivities. In general, the quality of GPR assessments will be better in hosts made of simple material (i.e., lossless, linear, isotropic, and homogeneous) with targets of high electrical contrast (i.e., ε, σ, and μ much different than those of the host medium), where high reflections occur [4,10].

1.3.3 Antennas

Antennas are responsible for the transmission of waves at the appropriate power levels and range of frequencies as well as for the reception of target signals. Their description is particularly useful because it provides a review of the possible configurations of

GPR systems, according to the specific intended application. The role of antennas is to transduce power, generated in the radio-frequency circuitry, propagated through a guiding structure (usually a transmission line), and coupled by the antenna to media surrounding the antenna (most often, but not always, air). In recent years, the demand for innovation in GPR antennas has been rapidly increasing due to the wider scope of applications covered by GPR. In the sense of performance characteristics, a high-quality antenna or array of antennas is essential for improvements in GPR resolution, penetration depth, and target detection. Additional requirements on antennas relate to the need for GPR to reduce the survey costs: low-profile, pure linear-polarization, and compact size, among others.

For instance, hydrological and geological applications often require high penetration depth with reduced resolution, whereas in other civil applications such as inspection of concrete, resolution is the primary requirement, whereas the depth of inspection is relatively low and either known or can be accurately estimated. Hydrological applications can do with larger size antennas which afford deeper penetration by operating at a lower range of microwave frequencies. On the other hand, concrete structures require compact size antennas that have higher resolutions but lower penetration depth. The choice of antennas establishes a trade-off between resolution and penetration depth according to the central frequency at which they operate.

An antenna can be modeled as a two-port transducer system characterized by a transfer function that can be frequency dependent and have a nonlinear phase response. As a consequence of this nonlinearity, the waveform that originates in the electronic circuitry will be significantly distorted, and the radiated pulse may be inappropriate for the post-processing stage. This fact can be critical for impulse wave radar, whereas in frequency-modulated or synthesized GPRs a simple calibration will fix the problem. In this sense, the choice between time and frequency-domain systems is another trade-off. Antennas for time-domain radars may produce ringing, signified by oscillations that can mask signals reflected from the targets. Frequency-domain radars require longer survey times and may be inappropriate for real-time applications.

In general, the antenna characteristics are influenced by the surrounding and nearby media. The efficiency of a GPR antenna can be substantially reduced when it operates in close proximity to a dispersive medium, not only because of the impedance mismatch caused by the medium but also because of the need for higher operating power which, in applications in isolated locations, may require larger batteries. Also, media in close proximity can affect the operational bandwidth of antennas. In general, most GPR antennas are wide bandwidth or UWB antennas in terms of the operational frequency range. As a rule, the higher the bandwidth the lower the efficiency, in terms of radiated power of the antenna. For this reason, the choice of antennas depends on the application. Borehole radars use narrower bandwidth antennas (such as dipole or loop antennas), whereas impulse radars use wideband or UWB antennas (bowtie, loaded bowtie, loaded dipole, spiral, horn, and Vivaldi types). Figure 1.3 illustrates two GPR antennas.

There are certain key antenna parameters that play particularly important roles in conjunction with GPR equipment [8]. First is the bandwidth of the antenna, described as the range of frequencies over which the antenna can maintain appropriate radiation

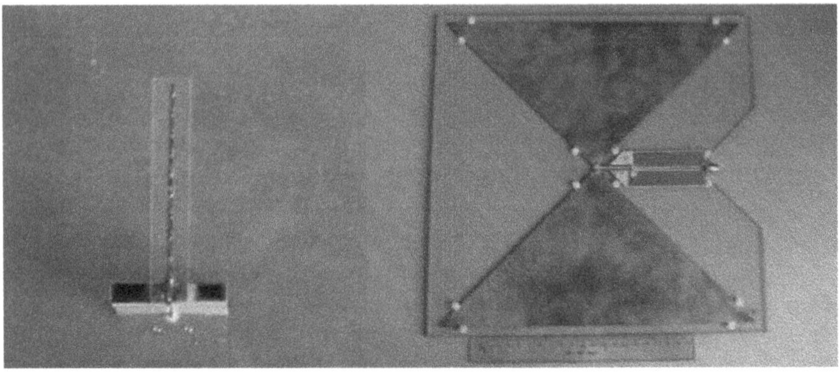

Figure 1.3 GPR antennas (loaded dipole on the left, planar bowtie on the right)

characteristics. Many antenna types have very narrow bandwidths and cannot be used to transmit pulses with wide frequency content (narrow pulses). Bandwidth is typically quoted in terms of the input impedance, although that is not the only criterion to be taken into account. A mismatch of the input impedance is due to the reflected waves within the antenna, which increase the voltage standing wave ratio.

Second, the polarization of the antenna plays a major role in the survey. Commercial GPR antennas can be found with linear and circular/elliptical polarization depending on the application so that polarization mismatches are minimized. As long as the polarization of the radiated wave can be affected by reflection/scattering phenomena or even by propagation in heterogenous soils, the use of elliptically polarized antennas seems advisable because they can receive both linear and ellipticaly (or circularly) polarized waves. Although the radiated power at a particular polarization is higher in a linearly polarized antenna, it may not be able to receive high-SNR signals from elliptically or circularly polarized antennas. Therefore, for targets oriented in a particular way (e.g., steel bars in concrete), the use of linear antennas could occasionally make sense. If a priori information on how the target is oriented is not available, elliptically polarized antennas are preferred. A good example of a practical antenna with elliptical polarization is the log-spiral (Figure 1.4). In this configuration, the transmitting antenna radiates two patterns that are orthogonal at any location in space. Using cross-polarized antennas, it is possible to detect simultaneously the target reflection in the two orthogonal polarizations [11], and some detected signal will be received no matter what the orientation of the target.

Additional attention should be paid to the directivity and gain of the antenna. The radiated power of any antenna varies in different directions in space, resulting in directive properties of the radiated fields. The directivity is mathematically defined as the ratio between the radiation intensity in a particular direction and the averaged radiation intensity, disregarding any losses that may have occurred in the antenna itself. If losses in the antenna are considered, by accounting for the input power available to the antenna, one defines the antenna gain or power gain. If no mention is made

Figure 1.4 Log-spiral antenna

of the direction associated with directivity, the maximum directivity or maximum gain should be understood, and manufacturers typically clarify in the characteristics sheet what is the direction of maximum radiation (further discussion of antennas and antenna parameters can be found in Chapter 3).

Finally, as a parameter that is particularly interesting for GPR purposes, we mention the antenna footprint. It is defined as the illuminated area in the horizontal plane at a given distance from the antenna, in the structure under test. This power projection is substantially influenced by the frequency, material properties, and antenna elevation. In general, when the material permittivity is higher the footprint is more compact, whereas for lower permittivities the footprints become larger. For closely spaced targets, a small footprint is desired since a small footprint helps in distinguishing between buried targets (higher power density at the targets). In addition, a small footprint is useful to reduce surface reflections and clutter. For optimal detection the footprint should comprise the size of the target or detection will be compromised. If the transmitted power of the antenna is spread across an area much larger than the target, then the signal backscattered will be much weaker than the signal reflected from the surface. The footprint has an elliptical shape as illustrated in Figure 1.5, with an approximate size [12]:

$$A = \frac{\lambda}{4} + \frac{d}{\sqrt{\varepsilon_r - 1}} \tag{1.12}$$

where A is the approximate long dimension radius of the footprint, λ is the wavelength of the EM wave, and d is the depth at which the footprint is sought. The optimal size and shape of the footprint can vary depending on the application. For instance, if the GPR is intended to detect long rebars in concrete structures, the footprint should have a longer ellipse shape, whereas small targets such as cracks require circular footprints.

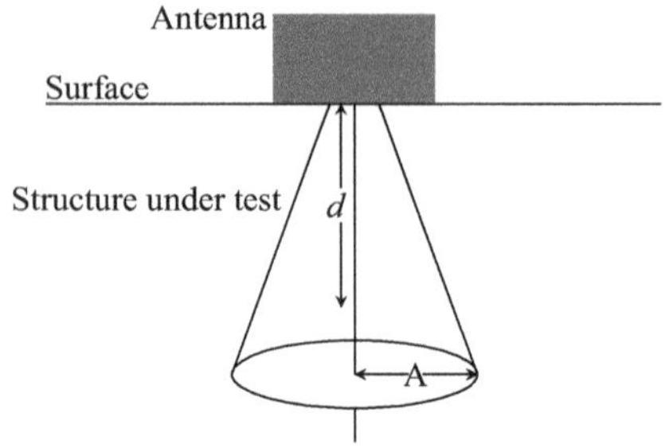

Figure 1.5 Antenna footprint

Table 1.2 GPR applications and choice of antennas

References	Application	Antenna type
[13–15]	Lossy dielectrics	Bowtie
[16–18]	Water leakage, metal ore detection	Loaded dipole
[19–21]	Forensic survey	Horn
[11,22]	Landmine, rebar detection	Spiral
[23,24]	Lossy, dispersive media	Vivaldi

Table 1.2 shows some applications of GPR and the different antenna types used for specific applications.

1.3.4 System specification

GPR systems can offer a variety of configurations depending on their system components. The system configuration directly affects the performance of the NDT assessment. Since some prior information about the structure under test and the target characteristics is often known, the system specification plays a major role in the success of the assessment.

In terms of particular applications, GPR can perform one or more of the following tasks:

1. the examination of a given volume in order to detect potential targets;
2. imaging of the target; and
3. classification of targets characteristics.

The criteria that should be evaluated for these purposes are the reliability of detection, accuracy of the predicted position/features of the target, resolution of the survey, and immunity to environmental noise. The reliability of detection should account for the detection range considering the expected features of the target and the host medium's electrical parameters, including possible non-homogeneities. To this end, the system must be able to distinguish clutter and noise signals, reflected signals from the surface, and real target signals in order to avoid false positives.

Target detection and classification should be done as accurately as possible and should include information about its position, size, shape, orientation, and electrical properties. As stated earlier, the host medium and target should have sufficiently different properties so that the incident wave can generate a strong reflected wave to be detected by the receiving antenna. In GPR surveys, targets must be resolved from each other by one or more spatial coordinates. This is called spatial resolution and measures the ability to distinguish (or discriminate) reflections from closely located targets.

Unintentional electromagnetic interferences can preclude or severely limit the use of GPR. For this reason, GPR systems should have low susceptibility (have high immunity) to environmental radio-frequency interferences. The RF immunity measures the ability of a GPR to perform its functions in the presence of undesired signals (noise). GPR can be classified by frequency of operation, and to minimize interference from external EM sources, GPR manufacturers are authorized by national government agencies to use specific bands of the EM spectrum.

The first choice to be made should be the frequency to be used in the survey, a choice that affects the most relevant features such as penetration depth and resolution. In fact, there is a trade-off in defining these two considerations. Since the depth of penetration is inversely proportional to attenuation, one can infer from (1.9) that higher frequencies will survey shallower depths than lower frequencies. On the other hand, from (1.10), a higher frequency f means a lower wavelength λ, which also implies a lower horizontal resolution (footprint) of the GPR system as predicted in (1.12) [10]. The frequency selection drives additional, important parameters in the GPR survey, including the sampling criteria in time and space. This is controlled by the acquisition parameters of the equipment, such as the time interval for sampling and the space between readings. To avoid aliasing, the Nyquist theorem provides maximum time and spatial intervals for a given frequency f as

$$\Delta t \leq \frac{1}{2f} \tag{1.13}$$

$$\Delta x \leq \frac{\lambda}{2} \tag{1.14}$$

In terms of vertical resolution (that is, the ability to discriminate stacked targets), the bandwidth of the system is relevant, and therefore, the radiated waveform is another choice to be made. As was mentioned previously, any EM pulse following a path through different media undergoes changes in its original waveform. These modifications will impact the GPR performance, and hence the waveform definition is a crucial factor. Some aspects of the waveform should be considered, including the following:

1. SNR;
2. avoidance of false positive detection;
3. resolution required; and
4. EM compatibility (EMC) issues.

It is important to note that these considerations are often in contradiction.

GPR can operate with a variety of modulation techniques. The most common are as follows:

- Continuous-wave (CW) radar utilizes stationary waveforms, usually sinusoids, instead of pulsed waveforms. This is an important issue in measuring range since there is no reference time between transmission and reception. In order to solve this problem, frequency-modulation CW (FMCW) of the waveform can be used. To avoid interruptions of the CW, two antennas are used in a transmitter/receiver configuration.
- Pulse-compression radar uses pulse waveforms coded with some modulation aimed at improving the signal bandwidth while providing similar vertical resolution to that of short pulses (in terms of temporal duration). In general, the modulation is frequency-based or phase-based. There are several types of pulse compression techniques for radar, with the chirp and stepped-frequency methods being the most common.
- Synthetic-aperture radar (SAR) maintains phase coherence for the transmitted pulse in a certain range of directions called antenna aperture. By analyzing the phase information of the received signal, a high-resolution image can be generated, thus revealing the features of the target. Large antenna apertures can be synthesized as superpositions of individual measurements made by small aperture antennas.
- Impulse radars were developed to emit very short pulses with high peak power. For this reason, the technique is often called short-pulse radar, impulse radar, or UWB radar.

Figure 1.6 shows the most common options for GPR systems in time and frequency-domain. Presently, most of the commercial systems use time-domain modulation, but the use of frequency-domain systems is growing fast due to the fact that they require less power and are more robust in the presence of noise [25,26].

Frequency-domain GPRs are based on band-limited signals. The most basic is the CW GPR, which transmits a signal of infinite duration in time and consequently narrowband in frequency, as a continuous sine wave, and receives reflections simultaneously. In this configuration, it is possible to detect buried targets but it is not possible to resolve range since the signals do not change with the range. For this reason, CW GPRs are not practical, and the signal bandwidth is widened by modulating

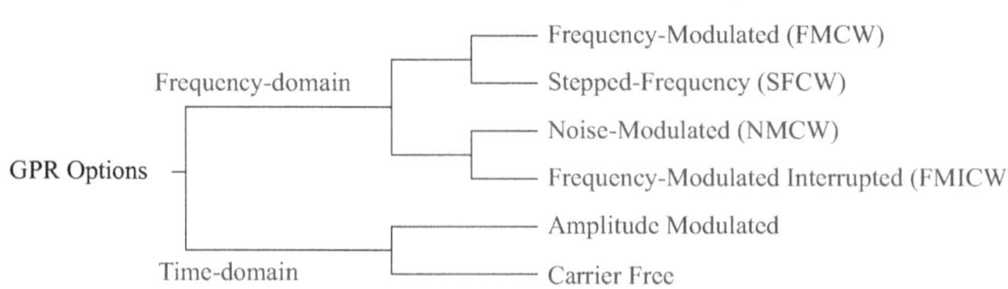

Figure 1.6 GPR systems and options of waveform modulation

it in a number of ways. A first choice is to employ amplitude modulation, in which the CW is modified by multiplying its amplitude by 1 or 0 (on and off states) at different times during the inspection. This option is commonly named pulsed GPR. Since the transmission and reception times are well defined by the equipment, it is possible to associate these times with the target range. However, for better resolution it is desirable to control the power spectral density of the pulse better than by the on/off method.

Another approach for modulation is to add more frequencies by increasing or decreasing the frequency of oscillation in the waveform (FMCW). Since the response from the target is frequency dependent, the survey is thereby improved. However, this type of GPR system suffers from interference issues. To reduce interference, a different method can be employed whereby the transmitted signal scans a certain interval of frequencies at varying frequency steps, covering thus the whole frequency range of the system in discrete steps. This technique is called SFCW. The use of frequency stepping avoids phase ambiguity by measuring the phase difference between the returned signals at each frequency. In addition, the hardware associated with SFCW is simpler than that employed in FMCW. A variation of the SFCW technique involves gating the timing of the transmitter and receiver circuits. It is possible to use other options, including noise-modulated continuous waveform (NMCW) [27] and frequency-modulated interrupted CW (FMICW) methods [28].

GPR systems that radiate time-limited waveforms and are aimed at detecting changes in their amplitudes are known as impulse GPRs. In these systems, the bandwidth is directly related to the pulse width. Shorter time pulses generate wider bandwidths and consequently require wideband antennas. Impulse GPRs are more prevalent than the FMCW and SFCW options due to the cost of RF components and hardware simplicity. However, they exhibit some drawbacks, including undesirable ringing, limited resolution, and phase group distortion.

Considering SNR, both FMCW and SFCW have lower noise figures and require less power to detect targets than impulse GPRs [11]. In terms of EMC issues, GPRs should avoid transmitting any spurious EM signals into the environment because these could interfere with other electronic systems. To this end, it is mandatory to comply with EMC standards adopted by national agencies such as the Federal Communications Commission in the United States and the European Telecommunications Standards Institute in Europe.

Another classification of GPR systems can be made in terms of the coupling to the structure under test, as air launched or ground coupled. Antennas can be in contact or even inside the host medium in ground-coupled GPRs, or they can be located above it in air-coupled GPRs. This choice will affect important features in the GPR assessment such as the footprint and the coupling signal, in amplitude and waveform, between the radiated pulse and that propagating in the host medium [11].

Table 1.3 shows some commercial GPR systems in terms of their coupling to the structure under test and frequencies of operation. The efficiency and input impedance of a GPR antenna can be substantially reduced when it operates close to a dispersive medium, requiring more power to operate. In such cases, the specification of how close to the structure under test the antenna can operate has to be taken into

Table 1.3 Commercial GPR systems

Manufacturer	Frequency [MHz]	Coupling
CyTerra	Not specified	Air
Ditch Witch	250, 700	Ground
Easyrad	20–400	Air
Easyrad	500–2,000	Ground
Geoscanners	210, 307, 380, 480, 675, 1,000, 1,500, 2,000	Ground
Logis-Geotech	250–700, 90–2,500	Ground
Logis-Geotech	400–2,500	Air
GSSI	270, 400, 900, 2,000	Ground
GSSI	1,600, 2,600	Air
IDS GeoRadar	400–900, 200–400, 600–900	Air
IDS GeoRadar	2,000	Ground
Mala Geoscience	200, 250, 1,200, 1,600, 2,300	Ground
TerraPlus	200, 450, 500, 750, 1,000, 1,600, 2,300	Ground
USradar	250, 500, 1,000, 1,500, 2,000, 2,500	Ground
3D-Radar	200–3,000	Ground

account. In general, ground-coupled GPRs can transmit/receive more power to the host medium, which results in clearer data and greater depth. But surface coupling and possible antenna ringing can result in difficulty in obtaining accurate predictions without appropriate signal processing. Air-coupled GPR measurements, on the other hand, must contend with the initial reflection from the air-ground surface and the reduced power being used in the survey. On the other hand, since antennas are not in contact with the ground, data can be collected at a much higher speed without risk of damage to antennas.

Once the desired waveform is generated, the transmitting antenna must be able to properly radiate it. This means the antenna should match the spectrum of the selected waveform. In this sense, a priori information about the test scenario is important in the selection process of an antenna structure to ensure it is less prone to undesired reflections, noise, and interference signals. Planar PCB antennas are often used in ground-coupled GPRs for their nondispersive characteristics. In laboratory tests and in controlled surveys, air-coupled GPRs using horn antennas have considerable advantages because of their narrow beamwidths and high gains over a wider frequency range.

The antenna structure can impact the GPR readings, imposing practical trade-offs. Available GPR equipment use a wide range of antennas with different physical sizes and radiation patterns, suited for various applications. Special attention has to be paid to the choice of the spatial interval of measurement Δx (see (1.14)), which could shorten the time of the survey at the cost of spatial resolution. A careful choice can avoid excessive volumes of data or spatially aliased responses. Improper choices can easily jeopardize the predictions obtained at the post-processing stage.

Recent years have brought new designs involving arrays of antennas, rather than the classical transmitter/receiver configuration. For bistatic (two antennas) or multistatic (array of antennas) cases, the distance between antennas should also be considered in the system specification. This feature will change the radiation pattern of the antenna and its coupling with the host medium, and it has to be made in accordance with the size of the target. The use of multiple-input–multiple-output array antenna technology has been proven to substantially improve the target detection resolution [29], and its use is expected to grow in the future.

In summary, the system specification defines how the GPR operates based on target characteristics and the structure under test. Important choices must be made, including the system bandwidth and the power level needed in the assessment. The definitions can improve the likelihood of detection producing better images. However, inaccurate EM modeling of the scenario can cause misinterpretation in direct readings, leading to wrong predictions in the actual location. In order to improve the data interpretation, PSTs are almost always necessary. These software tools substantially improve the assessment by eliminating noise and other potential sources of errors in the measurements.

1.4 Post-processing support tools

In addition to the concepts presented in the previous section, on-site GPR measurements require a post-processing stage to achieve the major goal of any prospection: prediction of geometrical features and physical properties of embedded objects in the host. This is key to addressing the many factors not discussed here and which affect the acquisition of real data: non-homogeneities of the host medium, lack of accurate material descriptions in actual scenarios, including varying environmental conditions, superposition of different scattering waves coming from multiple flaws, roughness of the interface of the host medium, possible EMC issues originating in external sources, changes in the footprint of the antenna due to the presence of nearby objects, and electrical noise in the equipment. These phenomena are a short list of major issues obviously neglected in the present basic introduction. As a result, EM wave propagation and scattering effects are so complex that any signal or image depicted from raw data provides only limited information, and PSTs constitute the only way to enhance the system performance providing concise predictions in real-world applications.

In most cases, GPR manufacturers provide proprietary software that take into consideration the particular system characteristics of the model and are aimed not only to provide survey maps of the site (generically named datagrams) but also to provide real-time checks of the quality of measurements. Any malfunction of the equipment (e.g., defective triggering pulses, inappropriate connections of cables) are also detected by this software, saving thus effort and time in prospecting.

Additional PSTs are available from various sources and may be applied to improve GPR surveys. In this context, it is useful to emphasize that generic codes which neglect the basic EM phenomena of Section 1.2 will be of limited utility in most cases, and it is only by taking into account these phenomena that the predictions start to be effective. In any case, there is still on-going research aimed to combine the latest techniques coming from signal and image processing theory with the specific EM properties of GPR systems [30]. Inclusion of advanced modeling features into PSTs increases successful rates, contributing thus to closing the so-called inverse problem gap which presently is the bottleneck in a broader acceptance of GPRs for use in generic applications. Table 1.4 shows a short list of available GPR software tools.

In short, the available information from GPR equipment after on-site measurements is far from being sufficient in providing the location and characteristics of the target. Auxiliary software tools may improve the interpretation of the collected data, thus reducing substantially the time and cost of the GPR detection. The objectives of these tools are to detect, locate, and define the target geometrical and physical properties. While basic techniques are focused on the first two goals, more advanced techniques deal with all objectives at once. The starting point to classify PSTs is based on how measurements are gathered. Based on the spatial acquisition of the survey, and recalling that a time-signal is measured for any on-site location, the available information can be processed separately at each location (called A-scans), or it may be grouped along lines (B-scans) and surfaces (C-scans) (Figure 1.7).

Figure 1.8 illustrates a C-scan as an assembly of B-scans. The A-scan results in one radar trace similar to any time-domain signal (Figure 1.7(a)). B-scans can be assembled by moving the GPR in one direction over the surface and generating A-scans at predefined spatial steps, producing datagrams depicted as plane views (Figure 1.7(b)). Finally, the compilation of data over a surface is called a C-scan, which is usually presented and processed as slices of B-scan datagrams. At this point, it is worth remarking that it is possible to acquire data at different heights over the surface. Although this approach can be considered strictly as a C-scan and is useful in dealing with the first wave reflected from the host or the varying footprint of the antennas, the common PSTs are not focused on this method of collecting data. In fact, the convention is to speak of signal processing techniques when A-scans are under consideration, and image processing techniques when B-scans are treated as a unique group. Regardless of the use of one or more of these techniques, a subsequent stage called pattern recognition is aimed at predicting accurately the position and character of targets based on signal or image analysis, or even soft-computing algorithms for learning and classifying purposes.

Table 1.4 A short list of GPR software tools

Software	**Application**
GPR Slice [31]	Archaeological and agriculture prospection
Radan [32]	Concrete, geology, forensics, and archaeological prospection
Radar Studio [33]	Archaeological, humanitarian, and environmental prospection
Ibis Guardian [34]	Geology prospection
GPRMax [35]	Electromagnetic simulation

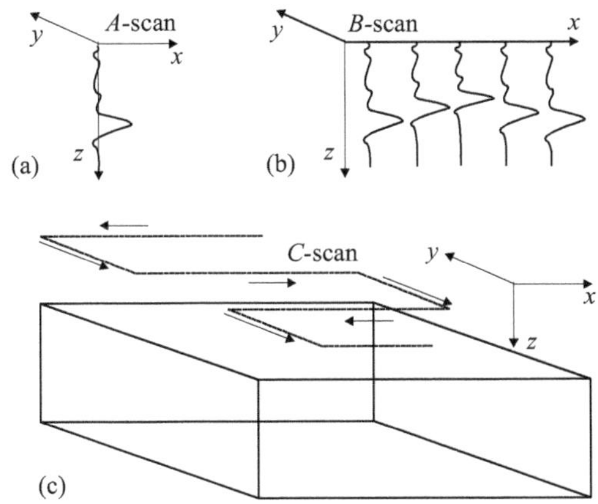

Figure 1.7 (a) A-, (b) B-, and (c) C-scans in a coordinate system

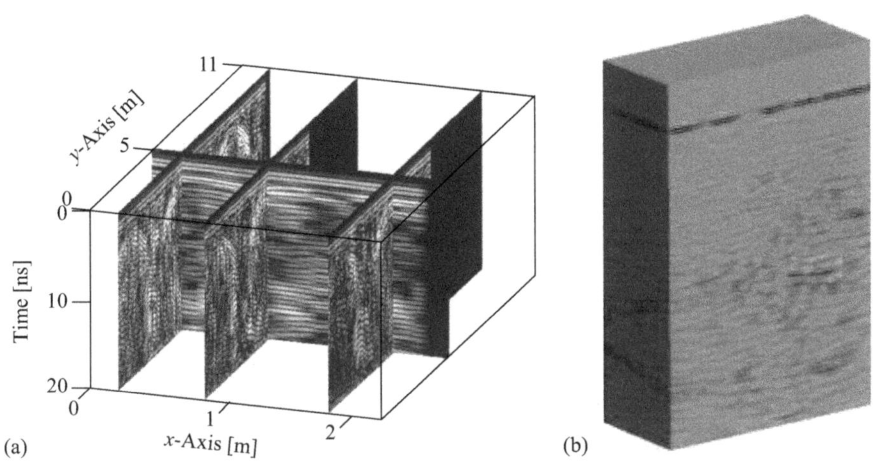

Figure 1.8 Generation of an image: (a) the superposition of B-scans and (b) 3D interpolation of C-scans to produce an image

1.4.1 Signal and image processing techniques

The overall objective of PSTs is to present information in a way that can be readily interpreted by the operator. To this end, processing and prediction procedures properly transform, separate, and classify the detected EM signal. The first step in the treatment of acquired signals is to distinguish between reflections coming from different targets, discarding those coming from objects that are different than the actual target, and then mapping the relevant reflections to form a new set of transformed signals or images ready to complete the prediction. Figure 1.9 shows an application of PSTs for the survey for marble at 900 MHz, displaying the predicted fractures.

(a)

(b)

Figure 1.9 (a) Marble block and C-scan of the block at 900 MHz and (b) predicted fractures

In general, the unwanted reflections come from early reflections from the air-structure interface or from late reflections from additional flaws in the host, which can appear as direct reflections from those objects or as secondary reflections (those appearing due to refraction or scattering of waves). Reflections originating at the air-structure interface usually vary with position, due to the varying heights of measurement on the surface (a very common effect in open-air fields where vegetation or rocks influence the survey), or from unmatched operation of the antenna due to the coupling with the host medium. Regarding late reflections, only relevant signals (i.e., those with an appreciable intensity when compared to those from the real target) need be taken into account. It is worth remarking that very different waveforms can appear depending on the characteristics of the reflection source (e.g., electrically small size objects versus heterogenous layered media), which challenges the PST ability to predict valid features. In general, all the processes mentioned here can be called clutter signals.

Clutter is recognized as one of the major limitations of the GPR technique, because poor signal-to-noise-clutter (SNC) ratios jeopardize the detection of the target. In some applications, for instance, the target may be buried at a shallow depth but the clutter is overspread on the target's response masking it. In addition, incoherent random noise may appear in heterogeneous soils as well as disturbances in readings coming from EM interferences from external sources. Another relevant issue arises when clutter signals are also frequency dependent, as is the case in lossy materials. The primary goal of signal processing techniques is to remove the signals interfering with the target's response.

In many instances, the SNC and SNR can be improved through simple strategies. For *A*-scans, the most basic procedures are the zero-offset removal, and the noise and clutter reduction based either on time or frequency-domain filtering. In some situations, a time-varying gain may improve the performance by enhancing late-time responses or offset removal. More elaborated strategies exist, including deconvolution techniques, spectral-analysis methods, or target resonances [4].

The one-dimensional information in an *A*-scan waveform contains not only the response from buried targets but also undesired signals, including the above mentioned first reflection from the surface, the receiver/transmitter EM interaction, thermal noise, interferences from other systems, and early reflections from shallow and medium-depth heterogeneities. The amplitude of the field reflected from the target is small compared to the undesired signals, but it can be extracted as long as the target and clutter responses are separated in time. Windowing and filtering techniques can be applied for this purpose. These can be considered as finite impulse response (FIR) filters, with a major advantage in terms of the linear phase response which preserves the original waveform of the field reflected from the target. Time-gating is the simplest method used to eliminate all unwanted signals in a time frame. However, the choice of the start and end time of the windowing is somewhat arbitrary, unless a priori physical parameters such as wave velocity and depth of the target are available (usually as estimates rather than exact values).

Since EM waves are attenuated with depth in the structure, most of the proprietary software provided by GPR manufacturers include time-variable gain processing tools. Using these, the gain can be adjusted with depth to amplify deeper reflections from possible targets, sometimes with customized gain functions. As a result, clutter and noise signals will be amplified as well. For this reason, this technique should be applied only when material properties are adequately assessed, otherwise, the time-varying gain will add dispersive behavior to the actual waveform with unpredictable consequences over the success rate of the pattern recognition stage. There are other techniques that can be applied in the time-domain, including simple mean scan subtraction, background removal, and moving average filtering [36,37] but, as with the time-varying gain, their use can increase distortion and/or reduce SNC and SNRs.

On the other hand, frequency-domain techniques offer an alternative outcome since they can select particular frequency sub-bands of interest in the GPR data. As for time-domain techniques, there are many possible strategies [36–38] for different purposes, commonly used to decrease some of the external interference signals or to remove noise. To decrease interference, a high-pass filter also known as dewow [4] can be used to decouple low-frequency interferences that can occur between the transmitter and receiver antennas [10]. Also in this group, noise appearing at intermediate frequency bands due to EMC issues can be extracted through notch filters [37], applying either finite impulse response (FIR) or infinite impulse response (IIR) filters. The second of these has a major negative effect in the form of nonlinear phase that tends to distort the waveform. Interference reduction requires longer computational time and is therefore not practical for large amounts of data.

For noise removal a low-pass filter can be applied, but this strategy usually subtracts some of the target response. The preferred method is to consider the received signal as a convolution of several impulse responses. The GPR assessment can be considered a problem whereby the input signal is a designed waveform that undergoes modification as it passes through a number of blocks before returning to the receiver

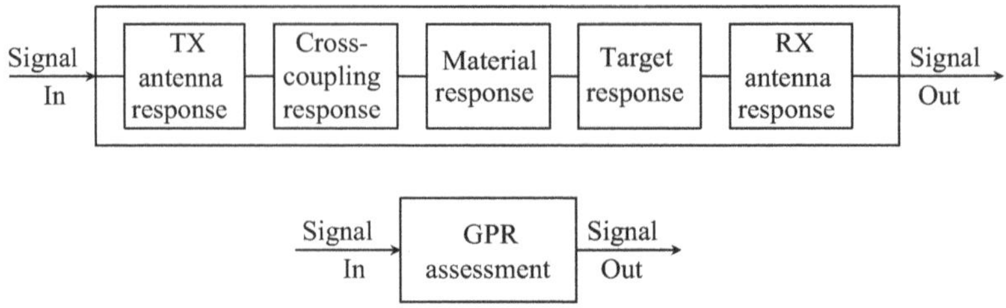

Figure 1.10 GPR transfer function

(Figure 1.10). The initial signal is then convolved with the impulse responses of each block (e.g., the antenna or the host medium), considered as boundaries where reflection/transmission phenomena arise. Deconvolution techniques aim to separate the signals appearing at each block and can be performed either in the space, time, or frequency-domain. Deconvolution is a time-consuming task in the time-domain, whereas in the frequency-domain it is performed by simple multiplications. When the Fourier transform needs to be applied to the data, it cannot be foreseen a priori which method is advantageous. Another step to improve the performance of the deconvolution techniques is to consider a wavelet decomposition instead of the Fourier transform. This opens new possibilities of filtering and amplifying waveforms aimed to increase the SNC and SNR.

A final mention to complete this short review of *A*-scan techniques is the use of resonances and other spectral-analysis strategies. The underlying idea in use of resonances is that every object possess unique resonant features calculated by Prony's method [39]. Through the knowledge of amplitudes and phases of the resonances, more efficient filters and/or amplifiers can be applied thus enhancing the performance of PSTs. Spectral-analysis methods, such as autoregressive moving average estimation or multiple signal classification, are usually applied in FMCW radars, where measured data are acquired directly in the frequency-domain.

B-scan signal processing techniques take advantage of the higher information achieved by spatial aggregation of measurements. It is important to realize that a common weak point of *A*-scan procedures is that they need to include a high response from the target, which usually happens when the GPR equipment is on top of the target. Obviously, in most scenarios these locations are unknown, and the first issue is how to detect locations of possible targets. To address this problem, the migration method may be used. The method is a deconvolution technique that allows refocusing of detected EM fields to their true spatial and temporal location [40]. Alternative approaches based on time and frequency-domain are, respectively, the Kirchhoff and frequency wave-number migration techniques [41,42]. On the other hand, synthetic-aperture processing combines the responses from the directive beams of antennas in *B*-scans to produce holographic SAR images [43]. Finally, *B*-scan information can be treated through image processing techniques [44], which have been proven to be effective in detecting edges of objects and in improving visual inspection through image filtering.

There is no technique that can work for all possible GPR configurations and applications. Some of them are not suitable for the GPR data available from specific surveys, or simply the resources needed for their implementation cannot be justified, or that the goals of the PST includes too many challenges. As an example of the latter case, the use of GPR for the detection and removal of landmines requires specific techniques aimed not only at the location of the target but also at providing sufficient information for safe extraction [11,45,46].

1.4.2 Pattern recognition

In GPR pattern recognition, a controlled training set of waveforms recorded from the receiving antenna is provided for a number of training scenarios. With this information, a mathematical or a theoretical procedure, usually involving soft-computing techniques (such as neural networks and machine learning), is derived as a predictive system. The quality of the predictive strategy can be further assessed for any attempt to predict the sources that created a new set of reflections coming from targets located in GPR scenarios not contained in the training set. The prediction of scattering sources through EM waves is a well-known ill-posed inverse problem, as opposed to the well-posed direct problems used to calculate the EM fields in physical scenarios where geometrical parameters or material properties are known. For NDE applications using GPR, pattern recognition aims at determining a useful finite number of parameters that characterize targets in a given medium by means of identification of their electrical and geometrical properties. In general, this is a challenging problem because the available data are often not sufficient for the prediction. This can usually be addressed through the insertion of some additional information of the GPR scenario, such as symmetries or other simplifying assumptions. On the other hand, in other cases a large number of data may be available but, because noise and other external sources can be present in the data, the information contains uncertainties which can bring the prediction system to unstable or inaccurate solutions.

The purpose of pattern recognition techniques is to improve the GPR assessment and diminish the importance of human interpretation by solving the inverse problem. In general, they are applied after signal processing techniques have been applied. To be effective, pattern recognition techniques should be cost effective, adaptable to different applications, provide reasonably fast response for large amounts of data, and provide a desirable false alarm response or probability of detection. Given the purpose of using GPR for NDT, we must assume that it is impossible to develop a technique that can classify and distinguish buried targets in all types of GPR assessments. There are, however, a number of applications that can be handled using kernel-based methods. A kernel function $k(x, x')$ characterizes the similarity between classes of objects [47]. In the GPR context, the objects are the received samples (time-domain) or patterns in the image (space-domain). Kernels are used in a number of applications as sketched in Figure 1.11.

Kernel-based models have two main advantages: one is freedom from local minima and the second is sparseness of the solution [48]. On the other hand, the choice of a kernel for a given application is a difficult task, as there are many types of kernels

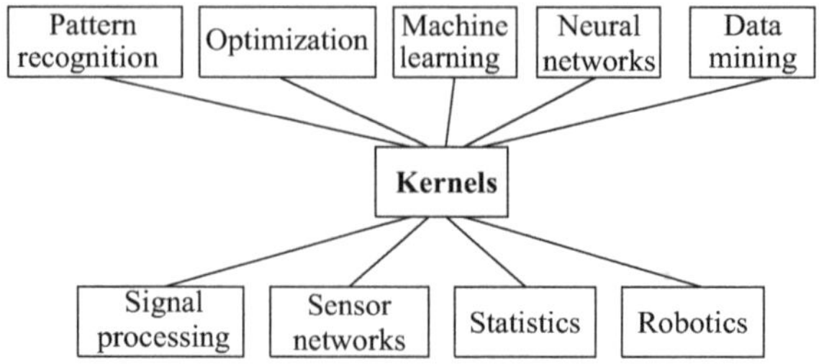

Figure 1.11 Applications that can be handled using kernel-based methods

Table 1.5 Examples of kernels [47]

Type of kernel	Function	Comments
Polynomial	$k(x,x') = \langle x,x'\rangle^d$	Homogeneous polynomial
Gaussian	$k(x,x') = \exp\left(\Vert x - x'\Vert^2/2\sigma^2\right)$	Gaussian radial basis
Sigmoid	$k(x,x') = \tanh(\kappa\langle x,x'\rangle + \vartheta)$	Gaussian with $\sigma > 0$
Triangular [49]	$k(x,x') = 1 - (\Vert x - y\Vert/\sigma)$	Euclidean distance

available. Some of them are shown in Table 1.5. Kernel-based methods also suffer from computational burden when implemented. That is due to the fact that Kernels operate on GPR data as though the data were projected onto a higher dimensional space.

In addition to the pattern recognition methods described in this subsection, other techniques, borrowed from a variety of disciplines, can be applied to GPR data, including but not limited to principal component analysis (PCA), discriminant analysis, Markov models, decision trees, k-nearest neighbors, edge histogram descriptors, spectral features, Bayesian classifiers, geometric features, texture features, neural networks, and support vector machines. Some of these are briefly discussed next.

1.4.2.1 Principal component analysis

In some situations, the dimension of the input vector is large, but the components of the vectors are highly correlated (redundant). PCA is a way of identifying patterns in data, and expressing the data in such a way as to highlight their similarities and differences [30]. The main advantage of PCA is that once these patterns have been identified, the data may be compressed by reducing the number of dimensions, without much loss of information. The technique is mostly used in image compression.

This technique affects the data as follows: it orthogonalizes the components of the input vectors (so that they are uncorrelated with each other), it orders the resulting orthogonal components (principal components) so that those with the largest variation come first, and it eliminates those components that contribute the least to the variation in the data set. The input vectors are first normalized so that they have zero mean and unity variance. For PCA to work properly, one has to subtract the mean from each of the data dimensions.

The PCA uses linear mapping of a given set of samples $S_q = \{x_1, \ldots, x_T | x_i \in \mathbb{R}^p\}$ to construct a new data set $S_p = \{y_1, \ldots, y_T | y_i \in \mathbb{R}^q\}$, where $p \leq q$. Considering a $p \times q$ V_{pq} matrix, the PCA algorithm can be described as follows:

1. subtract the mean from each data dimension,
2. calculate the covariance matrix,
3. calculate the eigenvectors and eigenvalues of the covariance matrix,
4. choose components and form a feature vector, and
5. derive the new data set.

Another interpretation of the PCA is the identification of directions that maximize the variance. The transformation V_{pq} generates a projection space in which the covariance matrix is diagonal. The diagonal covariance matrix implies that the variance of a variable with itself is maximized but is minimized with any other variable. Thus, the q variables with the highest variance in the new space should be kept. The principal components of a set of data in $\mathbb{R}^p$ provide a sequence of best linear approximations to that data, of all ranks $p \leq q$.

1.4.2.2 Discriminant analysis

Discriminant analysis is a pattern recognition technique that classifies the data, improving the mapping resulting from PCA. In general, it is applied after PCA has been applied to parameterize decision boundaries in a more effective way since it calculates the best discriminating components of the sample data. However, there is a trade-off to be considered in defining the PCA dimensional space prior to discriminant analysis in order to avoid the curse of dimensionality, an ill-conditioned problem [50], and a rise in the computational effort. This method usually fails for nonlinear GPR problems.

1.4.2.3 Feature selection

Feature selection is a process commonly used in machine learning, wherein a subset of the features available from the data is selected for application of the learning algorithm. It is different from feature extraction because the latter creates new features based on some combination of the original ones [51]. The best subset contains the least number of dimensions that most contribute to accuracy. The remaining unimportant dimensions are discarded. This is an important stage of preprocessing aimed at avoiding the well-known curse of dimensionality (another way is feature extraction) and, therefore, guarantee adequate convergence of the learning algorithm. It can also provide some understanding concerning the nature of the problem, as it indicates the main physical properties needed to classify an underground target. Therefore, feature selection is the task of choosing a small subset of features which is sufficient to predict the target and can capture the relevant information about the GPR problem.

The existing literature distinguishes between two types of feature selection algorithms: the so-called wrapper and filter methods [52]. Wrapper methods, which are computationally intensive and tend to over-fit [53], estimate the usefulness of a subset of features by a given predictor or learning algorithm. The wrapper methods try to directly optimize the performance of a given predictor. This is done by estimating the generalization performance. On the other hand, filter or variable ranking methods compute relevance scores for each feature and choose the most relevant ones according to those scores. This can be done using, usually in an ad hoc manner, evaluation functions aiming at searching the set of features that maximize the information. Among the commonly used evaluation functions are the mutual information, the margin, and dependence measures. The main drawback of such simple filtering methods is that they are not able to detect inter feature dependencies. One example of these dependencies in the GPR problem is the relation between clutter and antenna interference that can occur at the same time.

Features commonly found in GPR applications are spectral features, geometric features, and texture features. They can also be used in conjunction with other techniques such as neural networks or support vector machines [54,55].

1.4.2.4 Markov models

Markov models have been widely used in different fields of applications, including speech and handwriting recognition or to automatically align multiple biological sequences. The principle is based on modeling the statistical behavior of a time-varying parameter. In GPR, it can be used to automatically divide the received waveform into multiple sequence patterns. Those patterns can then be related to a database of possible targets by their signatures [56,57].

A hidden Markov model (HMM) consists of states and edges connecting the states. At each recorded sample time (or observation time), the process may be in one of these states. Since the actual state is not observable, it is called hidden. In this structure, there are transition probabilities between states that are connected by an edge. The HMM will be trained to establish the likelihood of each sample to be correlated to an objective. The GPR problem works in a similar fashion since data are recorded in time intervals over the structure under test. In addition, for this type of application, the states in HMM are finite. Furthermore, an HMM is called continuous if the state probability density functions are continuous and discrete if the state probability density functions are discrete [45]. HMM has proven to be a valuable tool for pattern recognition. However, a balance between important features in the scattered wave should be sought in order to improve the HMM performance [46].

1.5 Summary

This chapter has briefly discussed the use of GPR as an NDT and an NDE tool. In contrast to other NDT methods, GPR is based on EM wave propagation, which provides advantages and limitations. Among the advantages, one can cite relatively low-cost of survey, compact size of equipment, excellent trade-off between resolution and depth of detection, and suitability to an overwhelming variety of applications. However, the analysis of data is not straightforward not only because of the complexity of the physical phenomena involved in propagation and scattering of EM waves but also because of the impact of environmental conditions on the EM description of the media and consequently on the data acquired. To reduce this burden, the GPR equipment allows for multiple configurations that have to be carefully explored and selected in the planning stage of surveys. Even with the right choices of these configurations, the ability to predict and detect targets requires considerable expertise as well as a set of specialized post-processing tools. Far from being a closed line of research, the possibilities to improve detection are still open, and more exciting contributions are expected in the future to create more powerful NDT tools.

References

[1] Hulsenbeck & Co. German patent; 1926. No. 489434.

[2] Cook JC. Radar transparencies of mine and tunnel rocks. Geophysics, 40 (5); 1975:865–885. Available from: https://doi.org/10.1190/1.1440573.

[3] Evans S. Radio techniques for the measurement of ice thickness. Polar Record, 11 (73);1963:406–410.

[4] Daniels DJ. Ground Penetrating Radar. No. 2nd ed. in Electromagnetics and Radar Series. The Institution of Engineering and Technology, London; 2004.

[5] Yang CC and Bose NK. Landmine detection and classification with complex-valued hybrid neural network using scattering parameters dataset. IEEE Transactions on Neural Networks, 16 (3);2005:743–753.

[6] Donato LD and Crocco L. Model-based quantitative cross-borehole GPR imaging via virtual experiments. IEEE Transactions on Geoscience and Remote Sensing, 53 (8);2015:4178–4185.

[7] Du C, Zhang QL, Liu J and Peng M. Research of fusion method of GPR tomography images based on wavelet transform. In: 2016 16th International Conference on Ground Penetrating Radar (GPR); 2016:1–6.

[8] Balanis CA. Antenna Theory: Analysis and Design. v. 1. John Wiley & Sons, Hoboken, NJ, USA; 2005.

[9] Ida N. Engineering Electromagnetics, 4th ed. Springer, New York, NY, USA; 2020.

[10] Jol HM. Ground Penetrating Radar Theory and Applications. Elsevier Science, Amsterdam; 2008.

[11] Nicolaescu I. Improvement of stepped-frequency continuous wave ground-penetrating radar cross-range resolution. IEEE Transactions on Geoscience and Remote Sensing, 51 (1);2013:85–92.

[12] Neal A. Ground-penetrating radar and its use in sedimentology: Principles, problems and progress. Earth-Science Reviews, 66 (3);2004:261–330.

[13] Lestari AA, Bharata E, Suksmono AB, Kurniawan A, Yarovoy AG and Ligthart LP. A modified bow-tie antenna for improved pulse radiation. IEEE Transactions on Antennas and Propagation, 58 (7);2010:2184–2192.

[14] Sagnard F and Rejiba F. Wide band coplanar waveguide-fed bowtie slot antenna for a large range of ground penetrating radar applications. IET Microwaves, Antennas Propagation; 5 (6);2011:734–739.

[15] Giannakis I, Giannopoulos A and Warren C. A realistic FDTD numerical modeling framework of ground penetrating radar for landmine detection. IEEE Journal of Selected Topics in Applied Earth Observations and Remote Sensing, 9 (1);2016:37–51.

[16] Ebihara S, Hanaoka H, Okumura T and Wada Y. Interference criterion for coaxial-fed circular dipole array antenna in a borehole. IEEE Transactions on Geoscience and Remote Sensing, 50 (9);2012:3510–3526.

[17] Qin H and Xie X. Design and test of an improved dipole antenna for detecting enclosure structure defects by cross-hole GPR. IEEE Journal of Selected Topics in Applied Earth Observations and Remote Sensing, 9 (1);2016:108–114.

[18] Höfinghoff JF and Overmeyer L. Resistive loaded antenna for ground penetrating radar inside a bottom hole assembly. IEEE Transactions on Antennas and Propagation, 61 (12);2013:6201–6205.

[19] De Coster A, Tran AP and Lambot S. Fundamental analyses on layered media reconstruction using GPR and full-wave inversion in near-field conditions. IEEE Transactions on Geoscience and Remote Sensing, 54 (9);2016:5143–5158.

[20] Almeida ER, Porsani JL, Catapano I, Gennarelli G and Soldovieri F. Microwave tomography-enhanced GPR in forensic surveys: The case study of a tropical environment. IEEE Journal of Selected Topics in Applied Earth Observations and Remote Sensing, 9 (1);2016:115–124.

[21] Shao J, Fang G, Fan J, Ji Y and Yin H. TEM horn antenna loaded with absorbing material for GPR applications. IEEE Antennas and Wireless Propagation Letters, 13;2014:523–527.

[22] Liu H, Zhao J and Sato M. A hybrid dual-polarization GPR system for detection of linear objects. IEEE Antennas and Wireless Propagation Letters, 14;2015:317–320.

[23] Alkhalifeh K, Hislop G, Ozdemir NA and Craeye C. Efficient MoM simulation of 3-D antennas in the vicinity of the ground. IEEE Transactions on Antennas and Propagation. 64 (12);2016:5335–5344.

[24] Guan B, Ihamouten A, Dérobert X, Guilbert D, Lambot S and Villain G. Near-field full-waveform inversion of ground-penetrating radar data to monitor the water front in limestone. IEEE Journal of Selected Topics in Applied Earth Observations and Remote Sensing, 10 (10);2017:4328–4336.

[25] Pieraccini M. Noise performance comparison between continuous wave and stroboscopic pulse ground penetrating radar. IEEE Geoscience and Remote Sensing Letters, 15 (2);2018:222–226.

[26] Leckebusch J. Comparison of a stepped-frequency continuous wave and a pulsed GPR system. Archaeological Prospection, 18 (1);2011:15–25.

[27] Reeves B. Noise modulated GPR: Second generation technology. In: Proceedings of the 15th International Conference on Ground Penetrating Radar; 2014:708–713.

[28] Raimundo X, Salous S and Fioranelli F. Frequency modulated interrupted continuous wave signals in different radar imaging applications. In: 2014 XXXIth URSI General Assembly and Scientific Symposium (URSI GASS); 2014:1–4.

[29] Zeng Z, Li J, Huang L, Feng X and Liu F. Improving target detection accuracy based on multipolarization MIMO GPR. IEEE Transactions on Geoscience and Remote Sensing, 53 (1);2015:15–24.

[30] Rodriguez JB, Pantoja MF, Travassos XL, Vieira DAG and Saldanha RR. A prediction algorithm for data analysis in GPR-based surveys. Neurocomputing, 168;2015:464–474.

[31] Goodman. GPR Slice; 2018. Accessed: 2018-02-12. http://www.gpr-survey.com.

[32] GSSI. Radan Software; 2018. Accessed: 2018-02-12. https://www.geophysical.com/software.

[33] USRadar. Radar Studio; 2018. Accessed: 2018-02-12. http://www.usradar.com/ground-penetrating-radar-gpr-software/radar-studio/.

[34] GeoRadar. Ibis Guardian; 2018. Accessed: 2018-02-12. https://idsgeoradar.com/products/software/ibis-guardian.

[35] GPRMax. GPRMax; 2018. Accessed: 2018-02-12. http://www.gprmax.com.

[36] Haykin S and Van Veen B. Signal and Systems, 2nd ed. Wiley India Pvt. Limited, Hoboken, NJ, USA; 2007.

[37] Vetterli M, Kovačević J and Goyal VK. Foundations of Signal Processing. Cambridge University Press, Cambridge, UK; 2014.

[38] Madisetti V. The Digital Signal Processing Handbook. Electrical Engineering Handbook. Taylor & Francis, Boca Raton FL, USA; 1997.

[39] Blaricum MV and Mittra R. A technique for extracting the poles and residues of a system directly from its transient response. IEEE Transactions on Antennas and Propagation, 23 (6);1975:777–781.

[40] Leuschen CJ and Plumb RG. A matched-filter-based reverse-time migration algorithm for ground-penetrating radar data. IEEE Transactions on Geoscience and Remote Sensing, 39 (5);2001:929–936.

[41] Liu X, Serhir M, Kameni A, Lambert M and Pichon L. Ground penetrating radar data imaging via Kirchhoff migration method. In: 2017 International Applied Computational Electromagnetics Society Symposium – Italy (ACES); 2017:1–2.

[42] Sakamoto T, Sato T, Aubry P and Yarovoy A. Frequency-domain Kirchhoff migration for near-field radar imaging. In: 2015 IEEE Conference on Antenna Measurements Applications (CAMA); 2015:1–4.

[43] Skolnik MI. Introduction to Radar Systems. Electrical Engineering Series. McGraw-Hill, New York, NY, USA; 1980.

[44] González-Huici M. Accurate Ground Penetrating Radar Numerical Modeling for Automatic Detection and Recognition of Antipersonnel Landmines; 2013. Available from https://bonndoc.ulb.uni-bonn.de/xmlui/handle/20.500.11811/5789

[45] Gader PD and Mystkowski M. Applications of hidden Markov models to detecting land mines with ground-penetrating radar; 1999.

[46] Missaoui O, Frigui H and Gader P. Land-Mine detection with ground-penetrating radar using multistream discrete hidden Markov models. IEEE Transactions on Geoscience and Remote Sensing, 49 (6);2011:2080–2099.

[47] Scholkopf B and Smola AJ. Learning With Kernels: Support Vector Machines, Regularization, Optimization, and Beyond. MIT Press, Cambridge, MA, USA; 2001.

[48] Shawe-Taylor J and Cristianini N. Kernel Methods for Pattern Analysis. Kernel Methods for Pattern Analysis. Cambridge University Press, Cambridge, UK; 2004.

[49] Fleuret F and Sahbi H. Scale-Invariance of Support Vector Machines Based on the Triangular Kernel. In: 3rd International Workshop on Statistical and Computational Theories of Vision; 2003.

[50] Sun Y and Li J. Time-frequency analysis for plastic landmine detection via forward-looking ground penetrating radar. IEE Proceedings – Radar Sonar and Navigation, 150;2003.

[51] Dash M and Liu H. Dimensionality Reduction. John Wiley & Sons, Inc., Hoboken, NJ, USA; 2007.

[52] Jain A and Zongker D. Feature selection: Evaluation, application, and small sample performance. IEEE Transactions on Pattern Analysis and Machine Intelligence, 19 (2);1997:153–158.

[53] Křížek P, Kittler J and Hlaváč V. In: Kropatsch WG, Kampel M, and Hanbury A, editors. Improving Stability of Feature Selection Methods. Berlin, Heidelberg: Springer Berlin Heidelberg; 2007. p. 929–936.

[54] Kwak N and Choi CH. Input feature selection for classification problems. IEEE Transactions on Neural Networks, 13 (1);2002:143–159.

[55] Nguyen MH and de la Torre F. Optimal feature selection for support vector machines. Pattern Recognition, 43 (3);2010:584–591.

[56] Gader PD, Mystkowski M and Zhao Y. Landmine detection with ground penetrating radar using hidden Markov models. IEEE Transactions on Geoscience and Remote Sensing, 39 (6); 2001:1231–1244.

[57] Manandhar A, Torrione PA, Collins LM, and Morton KD. Multiple-instance hidden Markov model for GPR-based landmine detection. IEEE Transactions on Geoscience and Remote Sensing, 53 (4); 2015:1737–1745.

Chapter 2
Electromagnetic wave propagation

2.1 Introduction

The principles of ground penetrating radar (GPR) are solidly based on the use of electromagnetic wave radiation to detect buried targets. The present chapter attempts to describe the essentials of electromagnetic wave propagation and transport of power through space and through lossless and lossy materials. By necessity only the fundamental principles can be addressed in the context of this work and within the obvious space limitations. If further depth is required, the reader is encouraged to consult any of the many excellent books and other sources available, including some that are listed in this chapter as references, see, for example, [1–4] for basic theory and [5–10] for more advanced concepts.

To see how wave properties manifest themselves, consider first the motion of a tight string. Plucking a string, such as that of a guitar, it moves from side to side, generating a sound wave. There are, in fact, two effects here. One is the motion of the string itself, which is a wave motion. The second is the change in air pressure generated by the string's motion, which are the sound waves that are heard. The sound (frequency) depends on the size of the string (both length and thickness) and the amplitude depends on the displacement. The wave produced in the string in Figure 2.1, in the form of displacement of the string, y, is described by the following equation:

$$\frac{\partial^2 y}{\partial t^2} = \frac{Tg}{w}\frac{\partial^2 y}{\partial x^2} = v^2\frac{\partial^2 y}{\partial x^2} \tag{2.1}$$

where T is the tension in the string [N], g is the acceleration of gravity [m/s^2], and w is weight per unit length of the string [N/m]. The term Tg/w has units of [m^2/s^2] and is, therefore, a velocity squared. This is the velocity of propagation of the wave in the string. This equation is a scalar wave equation, and its solution should be familiar from physics as that of harmonic motion. The important point is that it defines the form of a wave equation; the function (displacement in this case) is both time and space dependent. There are other terms that may exist (such as a source term or a damping term), but the previous two terms are essential. The field so represented is a wave and has all the properties described earlier. Equation (2.1) is normally written in more convenient forms as

$$\frac{\partial^2 y}{\partial x^2} - \frac{1}{v^2}\frac{\partial^2 y}{\partial t^2} = 0 \tag{2.2}$$

The solution of (2.2) can be obtained in a number of ways. One is by separation of variables. The second is to introduce two new independent variables $\xi = x - vt$ and $\eta = x + vt$, substitute these into the wave equation, and perform the derivatives. Then, by integration on the two new variables, a general solution is obtained in the form

$$y(x,t) = g(x - vt) + f(x + vt) \tag{2.3}$$

where $g(x,t)$ and $f(x,t)$ are arbitrary functions that describe the shape of the wave. These may be the displacements of the string at any given time and location. For example, $\sin(x - vt)$ and $\cos(x - vt)$ may be appropriate functions. A better feel for what the solution means may be had by taking a very long string (such as a wire between two posts). The string is assumed to be infinite. A disturbance such as plucking the string at a time $t = 0$ is now created. This gives the initial condition $y(x,0) = g(x) + f(x)$. Consider, for example, the disturbance shown in Figure 2.2, created by moving the string as shown. If the string is let go, the disturbance moves in both directions at a velocity v. After a time t_1, the disturbances have moved to the right and left a distance vt_1 as shown. The disturbances propagating in the positive and negative x directions propagate away from the source and are called forward-propagating waves. For a vector field, such as the electric or magnetic field, an equation equivalent to (2.2) is the vector wave equation:

$$\frac{\partial^2 \mathbf{A}}{\partial z^2} - \frac{1}{v^2}\frac{\partial^2 \mathbf{A}}{\partial t^2} = 0 \tag{2.4}$$

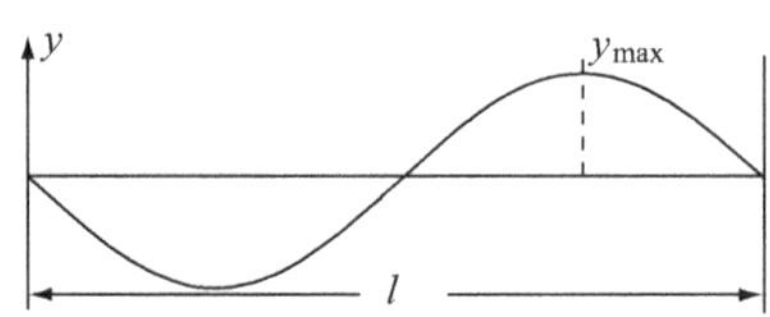

Figure 2.1 Wave motion of a tight string of length l

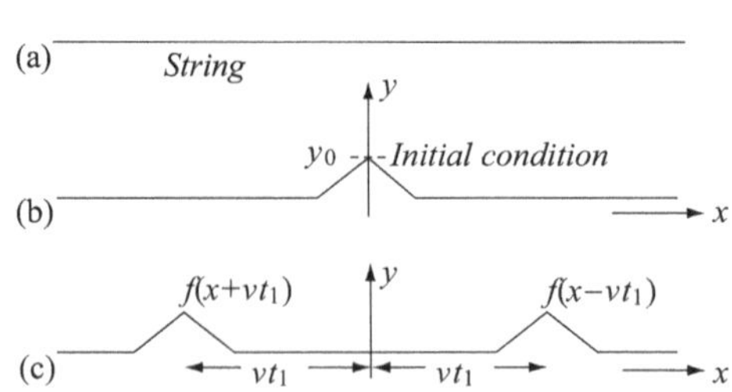

Figure 2.2 Propagation of a disturbance in a tight string: (a) string before the disturbance occurs, (b) the disturbance is introduced at $x = 0$, $t = 0$, and (c) the disturbance propagates in the positive and negative x directions at speed v

where **A** stands for any of the field vectors (**E**, **H**, etc.), v is the speed of propagation of the wave, and, in this case, the wave is assumed to propagate in the z direction (an arbitrary direction selected here for demonstration purposes).

The electromagnetic wave equation will be solved in the frequency-domain rather than in the time-domain, making use of the phasor notation. The propagation of the wave is real nonetheless. Speed of propagation, amplitudes, and all other aspects of the wave are similar to the previous simple problem, although the displacement of the string will be replaced by the amplitude of the electric or magnetic fields and the propagation will be in space (or, in the more general sense, in a medium that may be lossy or lossless). On the other hand, one can talk about propagation in a certain direction in space exactly like the propagation along the string, and a material-related velocity.

2.2 The electromagnetic wave equation and its solution

Based on the introduction of the displacement currents in Ampere's law, Maxwell predicted, on theoretical grounds, the existence of propagating waves, a prediction that was verified experimentally in 1888 by Heinrich Hertz. This prediction was based on the nature of the wave equations one obtains by using Maxwell's equations. The wave equation can be written in a number of forms, each useful under certain conditions. The solutions to the electromagnetic wave equations lead to a number of useful definitions, including phase velocity, wave impedance, and others, definitions to properties that are central to successful implementation of GPR systems.

Two types of equations will be discussed. One is the source-free wave equation, also called a homogeneous wave equation. The second is a complete equation, including source terms, and is called an inhomogeneous wave equation. The equations in the time-domain will be addressed first since pulsed GPR and behavior in nonlinear materials must be analyzed in the time-domain, but most of the work here will be in terms of phasors and the time-harmonic wave equation. It should also be remembered that homogeneity here relates to the form of the equation and should not be confused with material homogeneity.

The sections that follow look at the properties of propagating electromagnetic waves through the concept of plane waves. The proper wave equations are defined first from Maxwell's equations followed by a discussion of propagation in lossless and lossy media as well as the concepts of dispersion and polarization. These are followed by the encounter of electromagnetic waves with interfaces whereby the concepts of reflection, transmission, refraction, diffraction, and scattering are introduced.

2.2.1 The time-dependent wave equation

The electromagnetic wave equation is obtaining by starting with Maxwell's equations (see Section 1.3.1) and casting them in the form of (2.4) through simple substitution steps. Maxwell field equations in differential form are [1,2]

Faraday's law:

$$\nabla \times \mathbf{E} = -\frac{\partial \mathbf{B}}{\partial t} \tag{2.5}$$

Ampère's law:

$$\nabla \times \mathbf{H} = \mathbf{J} + \frac{\partial \mathbf{D}}{\partial t} \tag{2.6}$$

Gauss's law for the electric field:

$$\nabla \cdot \mathbf{D} = \rho_v \tag{2.7}$$

Gauss's law for the magnetic field:

$$\nabla \cdot \mathbf{B} = 0 \tag{2.8}$$

In addition, the material constitutive relations that are so critical to operation of GPR are

$$\mathbf{D} = \varepsilon \mathbf{E} \tag{2.9}$$

$$\mathbf{B} = \mu \mathbf{H} \tag{2.10}$$

Also, in conducting media

$$\mathbf{J} = \sigma \mathbf{E} \tag{2.11}$$

where ε [F/m], μ [H/m], and σ [S/m] are, respectively, the permittivity, permeability, and conductivity of the medium in which the waves propagate. As a reminder, **E** [V/m] and **H** [V/m] are called the electric and magnetic field intensities, respectively, whereas **D** [C/m^2] and **B** [T] are the electric and magnetic flux densities. **J** [A/m^2] is a current density that includes all possible types of currents, and ρ_v [C/m^3] is a volume charge density that includes all charge distributions.

Starting with (2.5), taking the curl on both sides:

$$\nabla \times (\nabla \times \mathbf{E}) = -\nabla \times \frac{\partial \mathbf{B}}{\partial t} = -\mu \frac{\partial (\nabla \times \mathbf{H})}{\partial t} \tag{2.12}$$

where (2.10) was used and, implicitly, the material was assumed to be linear. Substituting from (2.6) for $\nabla \times \mathbf{H}$:

$$\nabla \times (\nabla \times \mathbf{E}) = -\mu \frac{\partial}{\partial t}\left(\mathbf{J} + \frac{\partial \mathbf{D}}{\partial t}\right) = -\mu \frac{\partial^2 \mathbf{D}}{\partial t^2} \tag{2.13}$$

A more succinct form is obtained by using the identity $\nabla \times (\nabla \times \mathbf{E}) = -\nabla^2 \mathbf{E} + \nabla(\nabla \cdot \mathbf{E})$, substituting for **D** from (2.9) and rearranging the terms

$$\nabla^2 \mathbf{E} - \mu\varepsilon \frac{\partial^2 \mathbf{E}}{\partial t^2} = \mu \frac{\partial \mathbf{J}}{\partial t} + \nabla\left(\frac{\rho_v}{\varepsilon}\right) \tag{2.14}$$

In this form, the left-hand side is the same as the left-hand side of (2.4), whereas the right-hand side contains the sources of the wave in the form of time derivative of current densities and the gradient of volume charges densities. This is an inhomogeneous wave equation for the electric field intensity **E**. A similar equation may be obtained for the magnetic field intensity **H** by following an almost identical process. The homogeneous or source-free wave equation is therefore

$$\nabla^2\mathbf{E} - \mu\varepsilon\frac{\partial^2\mathbf{E}}{\partial t^2} = 0 \tag{2.15}$$

Although not explicitly shown here, the homogeneous wave equation for the magnetic field intensity is identical in the form of

$$\nabla^2\mathbf{H} - \mu\varepsilon\frac{\partial^2\mathbf{H}}{\partial t^2} = 0 \tag{2.16}$$

2.2.2 The time-harmonic wave equations

When the electromagnetic wave is monochromatic, such as in single-frequency continuous wave operation, both representation and analysis can be simplified using the phasor transformation to write Maxwell's equations and the resulting wave equation in the frequency-domain. There are two good reasons to do that in addition to the simplification afforded in spite of the fact that in GPR one is often called upon to analyze pulsed signals. First, any signal can be analyzed through superposition of monochromatic waves using the Fourier transform. Second, many of the commonly used terms such as wavelength, phase velocity, and wave impedance are only properly defined in the time-harmonic regime. In essence, the phasor transform replaces $\partial/\partial t$ by $j\omega$ and, implicitly, multiplies the amplitude by $e^{j\omega t}$. All fields and sources are now phasors (characterized by amplitude and phase) and solution is in terms of complex quantities.

Time-harmonic Maxwell's equations in (2.5)–(2.8) now become

$$\nabla \times \mathbf{E} = -j\omega\mathbf{B} \tag{2.17}$$

$$\nabla \times \mathbf{H} = \mathbf{J} + j\omega\mathbf{D} \tag{2.18}$$

$$\nabla \cdot \mathbf{D} = \rho_v \tag{2.19}$$

$$\nabla \cdot \mathbf{B} = 0 \tag{2.20}$$

The constitutive relations in (2.9)–(2.11) remain the same in form but now all quantities are phasors (although the same notation as in the time-domain is used, there is no room for confusion, since it is obvious from the relations that these are phasors).

The wave equation is obtained either by starting with the time-harmonic Maxwell's equations and following steps similar to those in the previous section, or with the time-dependent wave equation and then transforming to time-harmonic wave equations using the phasor transform. Using the latter (note that $\partial^2/\partial t^2$ converts to $-\omega^2$), the inhomogeneous wave equation becomes

$$\nabla^2\mathbf{E} + \omega^2\mu\varepsilon\mathbf{E} = j\omega\mu\mathbf{J} + \nabla\left(\frac{\rho_v}{\varepsilon}\right) \tag{2.21}$$

And the source-free wave equations in (2.15) and (2.16) become

$$\nabla^2\mathbf{E} + \omega^2\mu\varepsilon\mathbf{E} = 0 \tag{2.22}$$

$$\nabla^2\mathbf{H} + \omega^2\mu\varepsilon\mathbf{H} = 0 \tag{2.23}$$

2.2.3 The wave equation in lossy dielectrics

To understand the concept of propagation of waves in lossy media, it is useful to inspect the sources of the waves, either in the time-domain [Equation (2.14)] or in the frequency-domain [Equation (2.21)]. There are two sources shown. One is the gradient of the volume charge density (charge distribution). This source is of little or no consequence in GPR and therefore is disregarded here. However, the current density **J** is the sum of all possible sources of current. The most obvious is an applied current density, denoted as $\mathbf{J}_0$. However, in the GPR context, this source would imply that externally applied current sources exist in the medium being tested (soil). This does not happen, hence $\mathbf{J}_0$ is neglected here. But, in conducting media there may also be induced current densities and these may be denoted as $\mathbf{J}_e$. The latter obeys the constitutive relation in (2.11) and hence **J** can be replaced with $\sigma\mathbf{E}$. Equations (2.14) and (2.21) become (after removing the effect of charge densities)

$$\nabla^2\mathbf{E} - \mu\varepsilon\frac{\partial^2\mathbf{E}}{\partial t^2} = \mu\sigma\frac{\partial\mathbf{E}}{\partial t} \tag{2.24}$$

$$\nabla^2\mathbf{E} + \omega^2\mu\varepsilon\mathbf{E} = j\omega\mu\sigma\mathbf{E} \tag{2.25}$$

Equation (2.25) may be rewritten as follows:

$$\nabla^2\mathbf{E} + \omega^2\mu\varepsilon\left(1 - j\frac{\sigma}{\omega\varepsilon}\right)\mathbf{E} = 0 \tag{2.26}$$

Comparing (2.26) with (2.22) shows that they are the same if ε in (2.22) is replaced with $\varepsilon(1 - j\sigma/\omega\varepsilon)$. The latter term is called the complex permittivity, denoted by ε_c:

$$\varepsilon_c = \varepsilon' - j\varepsilon'' = \varepsilon\left(1 - j\frac{\sigma}{\omega\varepsilon}\right) \tag{2.27}$$

The concept of complex permittivity, in addition to being physically correct, (i.e., the permittivity of materials is, in general, a complex quantity dependent on frequency) can also be used to simplify the analysis of electromagnetic waves in that one can solve for the waves in lossless media and then substitute ε_c for ε to obtain the solution (as well as specific properties) in lossy media. This will be done in the following subsection. In addition, the complex permittivity in (2.27) shows that the higher the imaginary part, the higher the losses. Therefore, it also indicates how lossy a medium is through the ratio $\sigma/\omega\varepsilon$. The latter is called the loss tangent:

$$\tan\theta = \frac{\sigma}{\omega\varepsilon} \tag{2.28}$$

The loss tangent is a common way of specifying the lossy nature of materials.

2.2.4 Solution of the wave equation

Now that the electromagnetic wave equations have been obtained, it is the time to solve them. First, one must decide which wave equation to solve and under what conditions. In principle, it does not matter if one wave equation or another is solved, but, in practice, it is important to solve for the electric and magnetic fields in the domain of interest, in this case, in lossy and lossless media. To simplify things, the solution for the electric field intensity **E** and the magnetic field intensity **H** in lossless media and in the absence of sources is obtained first. The starting point is (2.22) or (2.26). To observe the behavior of fields and define the important aspects of propagation, it is simplest to use a one-dimensional wave equation; that is, to assume that the electric field intensity **E** or the magnetic field intensity **H** has a single component in space. The conditions under which the equations are solved are as follows:

1. Fields are time harmonic.
2. The electric field intensity is directed in an arbitrary direction (the x direction is used here) but varies in a direction perpendicular to the direction of the field (the z direction is used, arbitrarily); that is, the field is perpendicular to the direction of propagation.
3. The medium in which the wave propagates is lossless ($\sigma = 0$). Later, the permittivity will be replaced with the complex permittivity to obtain the solution in lossy media with no loss of generality.
4. The wave equation is source free ($\mathbf{J}_0 = 0$, $\rho_v = 0$).

This set of assumptions seems to be rather restrictive. In fact, it is not. Although the direction in space is fixed, one is free to choose this direction and can repeat the solution with a field in any other direction in space. Also, and perhaps more importantly, many of the previous assumptions are actually satisfied, at least partially in practice. For example, if the electric field intensity at the antenna of a receiver is needed, there is no need to take into account the actual current at the transmitting antenna: only the equivalent field in space. Similarly, propagation in general media, although not identical to propagation in lossless media, is quite similar in many cases. The benefit of this approach is in keeping the solution simple while still capturing all important properties of the wave. The alternative is a more general solution but one that is hopelessly complicated. Although it is not necessary to start with propagation in lossless media, this approach simplifies definitions of key properties of propagation. The addition of losses is then viewed as simply modifying the basic properties.

The conditions stated in this section specify what is called a uniform plane wave.

2.2.4.1 Solution for uniform plane waves

A uniform plane wave is a wave (i.e., a solution to the wave equation), in which the electric and magnetic field intensities are directed in fixed directions in space and are constant in magnitude and phase on planes perpendicular to the direction of propagation.

Clearly, for a field to be constant in amplitude and phase on infinite planes, the source must also be infinite in extent. In this sense, a true plane wave cannot be generated in practice. As a tool, or as an approximation, it has significant advantages. In numerical modeling, the plane wave is used for several purposes. For example, to test absorbing boundary conditions or to verify the radar cross-section of geometries. But it also serves as a local approximation of fields. As an obvious example, the wave generated by an antenna is certainly not a true plane wave. However, at a particular location in space or in a medium, one can reasonably assume that over a small enough area, the wave approximates a plane wave. The idea of plane waves also allows the definition of important quantities such as skin depth in a simple manner.

2.2.4.2 The one-dimensional wave equation in free-space and perfect dielectrics

With the assumptions in Section 2.2.4, the electric field intensity (in the Cartesian system of coordinates) is

$$\mathbf{E} = \hat{\mathbf{x}} E_x(z) \quad [\text{V/m}] \tag{2.29}$$

where $\mathbf{E}$ is phasor (i.e., $e^{j\omega t}$ is implied). These assumptions imply the following conditions: that $E_y = 0$, $E_z = 0$, $\partial E_* / \partial x = 0$, and $\partial E_* / \partial x = 0$, where * denotes any component of $\mathbf{E}$.

Substitution of these into (2.22) results in

$$\frac{d^2 E_x}{dz^2} + \omega^2 \mu \varepsilon E_x = 0 \tag{2.30}$$

where the partial derivative was replaced with the ordinary derivative because of the field dependence on z alone. Also, since the electric field is directed in a fixed direction in space, a scalar equation is sufficient. The term $\omega^2 \mu \varepsilon$ in (2.30) is denoted as k^2

$$k = \sqrt{\omega^2 \mu \varepsilon} = \omega \sqrt{\mu \varepsilon} \quad [\text{rad/m}] \tag{2.31}$$

Equation (2.30) is identical in form to (2.2) (except of course, it is written here in the time-harmonic form); therefore, it has the same type of solution. All that is needed is to find the functions $f(z)$ and $g(z)$ in the general solution in (2.3). In this case, since (2.30) describes simple harmonic motion, it has a solution

$$E_x(z) = E_0^+ e^{-jkz} + E_0^- e^{jkz} \quad [\text{V/m}] \tag{2.32}$$

where E_0^+ and E_0^- are constants to be determined from the boundary conditions of the problem. The notations (+) and (−) indicate that the first term is a propagating wave in the positive z direction called a forward-propagating wave and the second a propagating wave in the negative z direction called a backward-propagating wave, as in Figure 2.3 (horizontal arrows indicate the direction of propagation; the electric field intensity components are vertical). The amplitudes E_0^+ and E_0^- are assumed real (but they may, in general, be complex) and are arbitrary. This solution can be verified by direct substitution into (2.30).

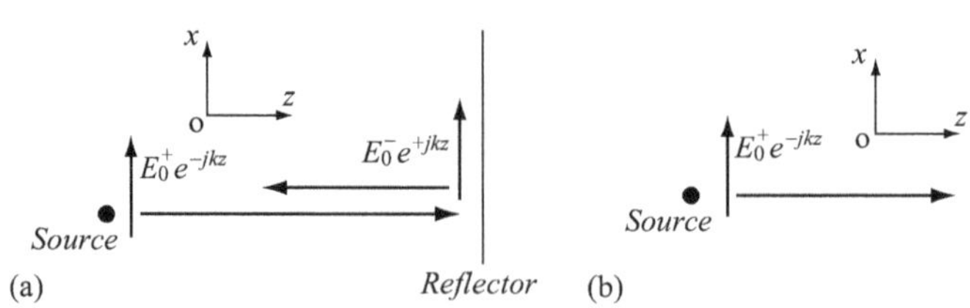

Figure 2.3 (a) Forward- and backward-propagating waves in bounded space and (b) forward-propagating wave in unbounded space (the horizontal arrows show the direction of propagation)

Using the phasor transformation, the solution in the time-domain may be written as

$$\begin{aligned} E_x(z,t) &= Re\left\{E_x(z)\,e^{j\omega t}\right\} \\ &= E_0^+\cos(\omega t - kz + \phi) + E_0^-\cos(\omega t + kz + \phi) \qquad [\text{V/m}] \end{aligned} \tag{2.33}$$

where the initial (arbitrary) phase angle ϕ was added for completeness.

If the wave propagates in boundless space, only an outward wave exists and E_0^- is zero (all power propagates away from the source and there can be no backward-propagating waves). If the forward-propagating wave is reflected without losses (i.e., for an electric field intensity, this means reflection by a perfect conductor), the amplitudes of the two waves are equal. Figure 2.3(b) shows schematically a forward-propagating wave without reflection. Assuming only a forward-propagating wave, the solution is

$$E_x(z) = E_0^+ e^{-jkz} e^{j\phi} \quad \text{or} \quad E_x(z,t) = E_0^+\cos(\omega t - kz + \phi) \qquad [\text{V/m}] \tag{2.34}$$

Examining these expressions, it becomes apparent that what changes with time is the phase of the wave. In other words, the phase of the wave travels at a certain velocity. To see what this velocity is, one can follow a fixed point on the wave in Figure 2.4, for which the phase of the field is $\omega t - kz + \phi = \text{constant}$:

$$z = \frac{\omega t}{k} + \frac{\phi}{k} - \text{constant} \tag{2.35}$$

The speed of propagation of the phase is

$$v_p = \frac{dz}{dt} = \frac{\omega}{k} = \frac{1}{\sqrt{\mu\varepsilon}} \qquad [\text{m/s}] \tag{2.36}$$

where $k = \omega\sqrt{\mu\varepsilon}$ was used. v_p is called the phase velocity of the wave. For a better feel for this velocity, think of a surfer catching a wave. The surfer rides the wave at a fixed point on the wave itself but moves forward at a given velocity. The surfer's velocity is equal to the phase velocity in the case of ocean waves. (The surfer is not moved forward by the wave; rather, the surfer slides down the wave. If it were not for this sliding, only bobbing up and down would occur as the phase of the wave moves forward.) In unbounded, lossless space, the phase velocity and the velocity of the wave or the velocity of transport of energy in the wave are the same.

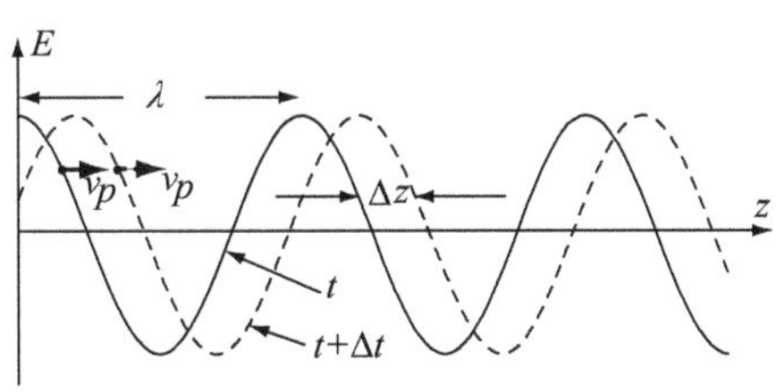

Figure 2.4 Definition of wavelength and calculation of phase velocity

The speed of propagation is a real speed, the speed at which energy propagates. The phase velocity is not a real speed in the sense that nothing material moves at that speed; only an imaginary point on the wave moves at this velocity. Because the phase velocity does not relate to physical motion, it can be smaller or larger than the speed of light and, as mentioned, may be different than the velocity of transport of energy. The phase velocity of electromagnetic waves is material dependent. In particular, in free space,

$$v_p = \frac{1}{\sqrt{\mu_0 \varepsilon_0}} = \frac{1}{\sqrt{4\pi \times 10^{-7} \times 8.854 \times 10^{-12}}} = 3 \times 10^8 \quad [\text{m/s}] \tag{2.37}$$

The phase velocity of electromagnetic waves in free space equals the speed of light, since light itself is an electromagnetic wave. The phase velocity in most materials is lower than c since $\mu_r, \varepsilon_r \geq 1$. In particular, the phase velocity in good conductors can be a small fraction of the speed of light in free space.

As the wave propagates, the distance between two successive crests of the wave depends both on the frequency of the wave and its phase velocity. The wavelength λ (in meters) is defined as that distance a wave front (a front of constant phase) travels in one cycle:

$$\lambda = \frac{v_p}{f} = \frac{2\pi v_p}{\omega} = \frac{2\pi}{\omega\sqrt{\mu\varepsilon}} = \frac{2\pi}{k} \quad [\text{m}] \tag{2.38}$$

Electromagnetic waves can be generated at any frequency, and hence the wavelength can vary from the very long (low frequency waves) to very short (high frequency waves). For example, the wavelength in free space for a wave at 50 Hz is 6,000 km. At 3 GHz (a frequency roughly in the middle of the range of frequencies used in GPR), the wavelength is 10 cm. From the definition of the wavelength in (2.38), k may be written as

$$k = \frac{2\pi}{\lambda} \quad [\text{rad/m}] \tag{2.39}$$

where k is called the wave number. If the wavelength in free space is given, then k is called the free-space wave number. However, inspection of (2.33) or (2.34) also shows k in the phase term of the solution. Hence, it is also called the phase constant. The term jk in (2.33) or (2.34) is called the propagation constant, since it is directly related to the phase velocity in (2.36). The propagation constant will be discussed in detail in the context of propagation in lossy media.

The same process can be performed for the magnetic field intensity **H**, since **E** and **H** obey identical equations [see (2.22) and (2.23)]. From the assumption that **E** has only an x component, which varies only in z, only the term $\partial E_x/\partial z$ exists. This means that $H_x = H_z = 0$ giving

$$H_y^+(z) = \frac{k}{\omega\mu} E_x^+(z) \quad [\text{A/m}] \tag{2.40}$$

This can be verified directly from Maxwell's equations. As was mentioned earlier, the reference field is **E** (an arbitrary choice used in electromagnetics as a convention).

It should also be noted that the ratio between $E_x^+(z)$ and $H_y^+(z)$ is a constant. This ratio is defined as

$$\eta = \frac{E_x^+(z)}{H_y^+(z)} = \frac{\omega\mu}{k} = \sqrt{\frac{\mu}{\varepsilon}} \quad [\Omega] \tag{2.41}$$

This quantity is an impedance because the electric field intensity is given in [V/m], and the magnetic field intensity is given in [A/m]. The quantity η is called the intrinsic impedance or wave impedance of the medium, since it is only dependent on material properties, as the right-hand side of (2.41) shows. The intrinsic impedance of free space is

$$\eta = \sqrt{\frac{\mu}{\varepsilon}} = \sqrt{\frac{4\pi \times 10^{-7}}{8.854 \times 10^{-12}}} = 377 \quad [\Omega] \tag{2.42}$$

Note that **H** and **E** propagate in the same direction and are orthogonal to each other and to the direction of propagation. This property makes **E** and **H** transverse electromagnetic waves. The relation between the electric and magnetic field intensities for plane waves in space is shown in Figure 2.5. This is a very special relation: for an electric field intensity in the positive x direction, the magnetic field intensity must be in the positive y direction (for the wave to propagate in the positive z direction), as the previous results indicate. This means that the wave transports power in the direction of the cross-product of **E** and **H** according to the Poynting vector: $\mathbf{P} = \mathbf{E} \times \mathbf{H}$ [W/m^2]. This is a property of plane waves but can also be seen in, for example, the fields of antennas, which are not strictly speaking plane waves (but can be approximated as such at large distances).

The discussion so far was restricted to a single component of the electric and magnetic field intensities. However, the same can be done with any other component of the electric or magnetic field and any other direction of propagation. The only real restriction on the previous properties was the use of the lossless wave equation. This will be relaxed later in this chapter when discussing the propagation of waves in materials.

The properties defined earlier are important properties of electromagnetic waves. These were defined for time-harmonic uniform plane waves, and, therefore, they are

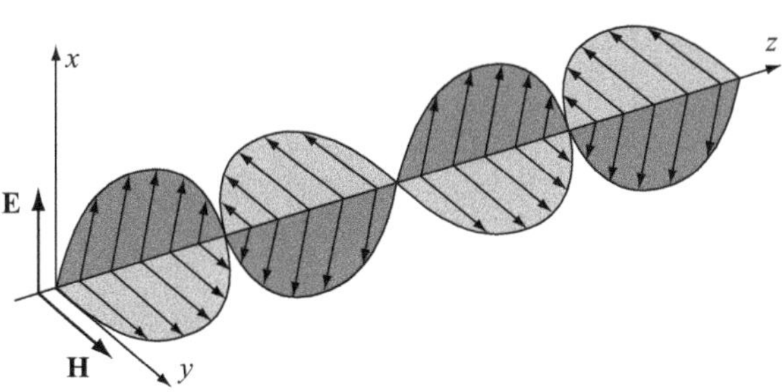

Figure 2.5 The relation between the electric and magnetic field intensities in a plane wave

only meaningful for time-harmonic fields. Wavelength and wave number can only properly be defined for time-harmonic fields. On the other hand, phase velocity and intrinsic impedance can be defined in terms of material properties alone and therefore do not depend on the time-harmonic form of the equations.

2.3 The electromagnetic spectrum

The previous section alluded to the fact that a low-frequency wave has a long wavelength and a high-frequency wave has a short wavelength, based on (2.38). This fact is quite important in applications of electromagnetic waves. Different portions of the spectrum of electromagnetic waves serve for different applications. For example, frequency-modulated (FM) radio transmission in the United States is between 87.5 and 108 MHz, whereas most satellite communication occurs between 1,000 MHz (1 GHz) and 30,000 MHz (30 GHz). Similarly, cell phones may be operating in the 1.9 GHz band (1,850–1,990 GHz) or in the 800 MHz band (824–894 MHz). These allocations may be arbitrary (allocated by convention or by an agency), or may be based on traditional use, on availability of spectrum or, more often on the propagation properties of the wave in a given band of frequencies to ensure optimal operation. The electromagnetic spectrum is divided into bands, based either on frequencies or the equivalent wavelength in free space. These bands are, to a large extent, arbitrary and are designated for identification purposes. A simplified graphical representation of the electromagnetic spectrum is shown in Figure 2.6. The following should be noted:

1. The spectrum of electromagnetic waves is between zero and ∞. Although it may not be practical (or possible) to use waves above certain frequencies, they can and do exist.
2. Infrared, visible light, ultraviolet rays, X-rays, α, β, and γ-rays, cosmic rays, etc. are all electromagnetic waves.
3. The narrower bands below the infrared band are arbitrarily divided by wavelength and designated names. Each band is one decade in wavelength.

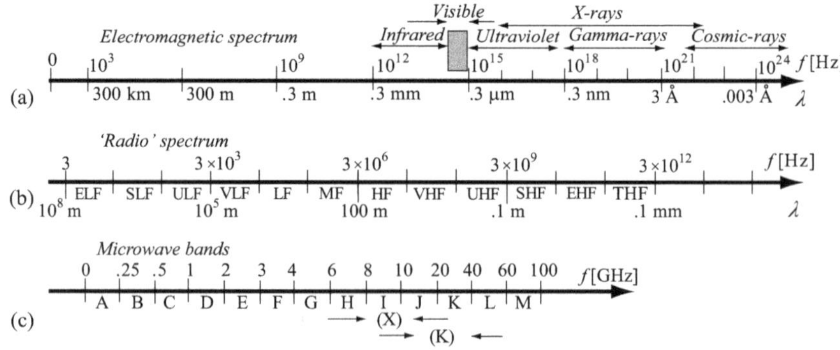

Figure 2.6 The electromagnetic spectrum: (a) general, (b) radio-frequency bands, (c) microwave bands

Most of our work will have to do with the spectrum below the infrared region, since much of the work on light is treated in optics. However, some of the relations usually associated with optics (Snell's laws, reflection, transmission, and refraction of waves) apply equally well to lower frequencies.

Use of the electromagnetic spectrum is based on the needs of the various applications and these do not follow any particular, designated band. For example, radar, as used for aircraft detection, guidance, and weather, operates in the super-high frequency (SHF) and extremely-high frequency (EHF) bands. GPR applications typically span from a few hundred kHz to a few GHz. As such they operate in the medium frequency (MF) through the super-high frequency (SHF) bands. The frequency used in any particular application has important consequences on performance. In radar equipment, the higher the frequency, the higher the resolution. In communication with satellites, the size of the antennas is dictated by frequency (the higher the frequency, the smaller the antenna). It is therefore of some advantage to use higher frequencies. In communication with submarines, the main effect is that of penetration of waves in water. Low frequencies penetrate well, whereas high frequencies do not. Similarly, residential microwave ovens operate mostly at 2,450 MHz because at that frequency water absorbs electromagnetic energy and can be heated.

The spectrum may be further subdivided for specific purposes. For example, the very-high frequency (VHF) band is divided to allocate frequency bands (each 6 MHz wide) to TV channels. Again, this is by convention. Similarly, the microwave region is often divided in bands, each designated with a letter as shown in Figure 2.6(c). In this definition, microwave ovens operate in the E-band and police radar detectors operate in the I-band (previously known as the X-band; microwave bands shown in brackets are old designations shown here for comparison) or K-band.

2.4 Propagation of plane waves in materials

That waves are affected by the material in which they propagate has been shown earlier, where propagation in lossless dielectrics, including free space, was discussed. The phase velocity, wavelength, wave number, and intrinsic impedance are material dependent. It is also known from day-to-day experience that different materials affect waves differently. For example, when passing under a bridge or through a tunnel, the car radio ceases to receive. Propagation in water is vastly different than propagation in free space. In listening to a shortwave radio, one experienced much better reception during the night than during the day. All these are due to effects of materials or environmental conditions on waves.

This aspect of propagation of waves is discussed next because it is extremely important both to understanding of propagation and the use of electromagnetic waves in GPR as well as in other applications. Based on the propagation properties of waves, one can choose the appropriate frequencies, type of wave, power, and other parameters needed for design. GPR relies on reflections of waves from targets and these reflections depend on properties of the targets and the surrounding media. The depth to which GPR is useful is directly related to losses and material properties, and variations in them affect the contrast in any GPR survey.

It is therefore essential to define the important parameters of propagating waves in general, lossy media. These parameters include the propagation, phase, and attenuation constants, as well as the complex permittivity (see (2.27)) and the skin depth (see (2.59)). These parameters will then be used for the remainder of the book to describe the behavior of waves in a number of important configurations, including antennas and the modeling of the GPR assessment environment.

2.4.1 Propagation of plane waves in lossy dielectrics

A lossy dielectric is a material that, in addition to polarization of charges, conducts free charges to some extent. In simple terms, it is a poor insulator, whereas a perfect dielectric is a perfect insulator. For the purpose here, a lossy dielectric is characterized by its permeability, permittivity, and conductivity. Thus, in addition to displacement currents, there are also conduction currents in the dielectric. The assumption that there are no sources (free currents and charge distributions) in the solution domain is still valid.

The approach here is to use the equations and parameters obtained for the lossless dielectrics and replace the permittivity ε with the complex permittivity ε_c (see (2.27)). This provides the correct relations for lossy dielectrics. However, not all lossy dielectrics are the same. In some, the losses are very low, whereas in others, they may be very high, resembling conductors. These extremes will be analyzed using the loss tangent in (2.28), resulting in approximations that may be used for very low and very high loss media.

The starting relation is the source-free one-dimensional equation [Equation (2.30)] in which ε is replaced with the complex permittivity ε_c to get:

$$\frac{d^2E_x}{dz^2} + \omega^2\mu\varepsilon\left(1 - j\frac{\sigma}{\omega\varepsilon}\right)E_x = 0 \tag{2.43}$$

Comparing this with (2.30), ε in (2.31) can be replaced with ε_c to obtain k_c, the lossy equivalent to k:

$$k_c = \omega\sqrt{\mu\varepsilon\left(1 - j\frac{\sigma}{\omega\varepsilon}\right)} = \omega\sqrt{\mu\varepsilon}\sqrt{\left(1 - j\frac{\sigma}{\omega\varepsilon}\right)} \tag{2.44}$$

This is a complex term and hence the units have been removed for now. They will be reintroduced shortly. Since the solution for the lossless wave equation is known, (given in (2.33)), the solution for the lossy case is

$$E_x(z) = E_0^+e^{-jk_cz} + E_0^-e^{jk_cz} \qquad [\text{V/m}] \tag{2.45}$$

where jk_c is called the (complex) propagation constant and may be written as

$$\gamma = jk_c = j\omega\sqrt{\mu\varepsilon}\sqrt{\left(1 - j\frac{\sigma}{\omega\varepsilon}\right)} \tag{2.46}$$

since γ is a complex number, it may be written formally as $\gamma = \alpha + j\beta$ and (2.45) rearranged as

$$\begin{aligned} E_x(z) &= E_0^+ e^{-(\alpha+j\beta)z} + E_0^- e^{(\alpha+j\beta)z} \\ &= E_0^+ e^{-\alpha z} e^{-j\beta z} + E_0^- e^{\alpha z} e^{j\beta z} \qquad \text{[V/m]} \end{aligned} \tag{2.47}$$

To better see the significance of the propagation constant, it is useful to rewrite (2.47) in the time-domain:

$$E_x(z) = E_0^+ e^{-\alpha z} \cos(\omega t - \beta z) + E_0^- e^{\alpha z} \cos(\omega t + \beta z) \qquad \text{[V/m]} \tag{2.48}$$

This form shows explicitly that the amplitude E_0^+ attenuates exponentially as the wave propagates in the positive z direction, whereas E_0^- attenuates in identical fashion as the wave propagates in the negative z direction. The term α is therefore called the attenuation constant. Similarly, the term β delays the phase as the wave propagates. Therefore, it is called the phase constant. The attenuation and phase constants are found directly from (2.46) as

$$\alpha = \omega \left[\frac{\mu\varepsilon}{2} \left(\sqrt{1 + \left(\frac{\sigma}{\omega\varepsilon} \right)^2} - 1 \right) \right]^{1/2} \qquad \text{[Np/m]} \tag{2.49}$$

measured in nepers/meter [Np/m], and the velocity of propagation will be determined through the phase constant β:

$$\beta = \omega \left[\frac{\mu\varepsilon}{2} \left(\sqrt{1 + \left(\frac{\sigma}{\omega\varepsilon} \right)^2} + 1 \right) \right]^{1/2} \qquad \text{[rad/m]} \tag{2.50}$$

measured in rad/m.

The attenuation constant α is measured in Np/m. The neper is a dimensionless constant and defines the fraction of the attenuation the wave undergoes in 1 m. Attenuation of 1 Np/m reduces the wave amplitude to $1/e$ as it propagates a distance of 1 m (e is the Euler number and equals approximately 2.71828). Therefore, it is equivalent to 8.69 dB/m ($20\log_{10}e = 8.69$), that is, 1 Np/m = 8.69 dB/m.

Figure 2.7 shows schematically the behavior of a forward-propagating wave with attenuation α.

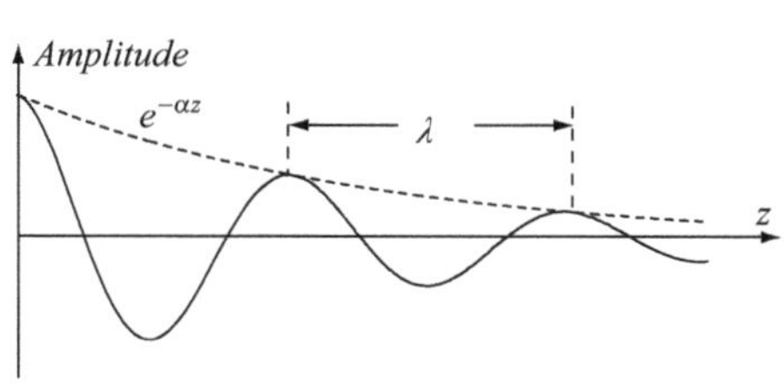

Figure 2.7 Propagation of a wave in a lossy material showing exponential attenuation

The other quantities, including phase velocity, wavelength, and wave impedance, change as well. Phase velocity is $v_p = \omega/\beta$ hence

$$v_p = \frac{\omega}{\beta} = \frac{1}{\sqrt{\frac{\mu\varepsilon}{2}\left[\sqrt{1+\left(\frac{\sigma}{\omega\varepsilon}\right)^2}+1\right]}} \quad [\text{m/s}] \tag{2.51}$$

The wavelength is given as $\lambda = 2\pi/\beta$:

$$\lambda = \frac{2\pi}{\beta} = \frac{2\pi}{\omega\sqrt{\frac{\mu\varepsilon}{2}\left[\sqrt{1+\left(\frac{\sigma}{\omega\varepsilon}\right)^2}+1\right]}} \quad [\text{m}] \tag{2.52}$$

Similarly, substituting the complex permittivity in (2.41) results in the wave impedance for lossy media:

$$\eta = \frac{j\omega\mu}{\gamma} = \frac{j\omega\mu}{\sqrt{j\omega\mu(\sigma + j\omega\varepsilon)}} = \sqrt{\frac{j\omega\mu}{\sigma + j\omega\varepsilon}} \quad [\Omega] \tag{2.53}$$

Inspection of (2.45)–(2.53) reveals first that these are essentially of the same form as for the lossless medium once the complex permittivity is used but also that the phase velocity and wavelength are smaller than for a perfect dielectric with the same permittivity and permeability since the phase constant β is larger than k. The wave impedance is now complex and decreases in magnitude as the conductivity increases. The only term that did not exist in perfect dielectrics is the attenuation constant. These properties play crucial roles in the application of GPR; in broad terms, the attenuation constant is responsible for the depth of detection, whereas the impedance contrast between targets and background medium is responsible for sensitivity of detection.

In most cases encountered in GPR work, the general lossy relations in this sections apply, that is, one has to view materials as general and no particular assumptions may be imposed on the losses in them. There are, however, situations in which the losses are either very low or very high. Under these conditions, the expressions and therefore the analysis can be simplified. A medium is considered low loss if the loss tangent is low, specifically, if $\sigma/\omega\varepsilon \ll 1$ (see (2.27) and (2.28)). Nevertheless, one cannot simply neglect this term in the complex permittivity because that would imply neglecting losses altogether. Rather, the propagation constant (2.46) is expanded using the binomial expansion and the first complex term is retained as approximations to the attenuation and phase constants:

$$\gamma = j\omega\sqrt{\mu\varepsilon}\sqrt{\left(1 - \frac{j\sigma}{\omega\varepsilon}\right)} \approx j\omega\sqrt{\mu\varepsilon}\left(1 - \frac{j\sigma}{2\omega\varepsilon} + \text{Higher Order Terms}\right) \tag{2.54}$$

This provides approximations for the attenuation and phase constants for low-loss dielectrics as

$$\alpha \approx \frac{\sigma}{2}\sqrt{\frac{\mu}{\varepsilon}} = \frac{\sigma}{2}\eta_n \quad [\text{Np/m}], \qquad \beta \approx \omega\sqrt{\mu\varepsilon} \quad [\text{rad/m}] \tag{2.55}$$

where η_n is the wave impedance of the equivalent lossless medium (i.e., a medium with the same permittivity and permeability but no losses).

It should be noted that under these conditions, the changes in phase velocity and wavelength are neglected, the attenuation constant is finite (but small), and the wave impedance has a small imaginary part and a much larger real part. In some cases, the change in the wave impedance is negligible and the lossless wave impedance may be used. This latter approach is common in GPR, whereby low loss media may be assumed to be lossless simply because the exact conductivity is now known.

High loss media are closer to true conductors than to dielectrics, and one can expect their behavior to be different. Here again, the approximations afforded are linked to the complex permittivity in (2.27) and the loss tangent in (2.28), namely, one may now assume $\sigma/\omega\varepsilon \gg 1$ and write the propagation constant as

$$\gamma \approx j\omega\sqrt{\mu\varepsilon}\sqrt{-\frac{j\sigma}{\omega\varepsilon}} = (1+j)\sqrt{\pi f \mu\sigma} \tag{2.56}$$

The attenuation and phase constants can now be approximated as

$$\alpha \approx \sqrt{\pi f \mu\sigma} \quad [\text{Np/m}], \qquad \beta \approx \sqrt{\pi f \mu\sigma} \quad [\text{rad/m}] \tag{2.57}$$

The changes in phase velocity, wavelength, and wave impedance are now drastic. These are approximated as

$$v_p \approx \omega\delta \quad [\text{m/s}], \qquad \lambda \approx 2\pi\delta \quad [\text{m}], \qquad \eta \approx (1+j)\pi f \mu\delta \quad [\Omega] \tag{2.58}$$

where

$$\delta = \frac{1}{\alpha} = \sqrt{\frac{1}{\pi f \mu\sigma}} \quad [\text{m}] \tag{2.59}$$

is called the skin depth or the standard depth of penetration. The skin depth is simply a measure of the attenuation, and it indicates that the quantity of interest, say the electric field intensity attenuates by a factor of $1/e$ in one skin depth (see the definition of the neper following (2.50)). The skin depth affords a simple measure of estimating the amplitude of a wave with depth. As an example, the amplitude of a wave propagating to a depth of 5 skin depths reduces to about 0.1% of its amplitude at the surface. Skin depth is associated with high loss dielectrics, and its use as an estimator for depth should be used carefully because its definition assumes plane waves.

Some important general observations are appropriate here by way of summary:

1. The phase velocity in lossy dielectrics is lower than in perfect dielectrics. This can be seen from (2.51), since β for lossy materials is larger than k for perfect dielectrics. The larger the losses, the lower the phase velocity. The same applies to wavelength.
2. The intrinsic impedance (wave impedance) in lossy dielectrics is complex, indicating a phase difference between the electric and magnetic field intensities in the same way as the phase difference between voltage and current in a circuit that contains reactive components. The magnitude of the intrinsic impedance is

lower in conductive media. The higher the conductivity (losses), the lower the magnitude of the impedance.

3. The electric and magnetic field intensities remain perpendicular to each other and to the direction of propagation regardless of losses. This is a property of the uniform plane waves and attenuation does not change that.
4. Attenuation of the wave in lossy media is exponential. This means that in materials with high conductivity (high loss), the attenuation is rapid.

2.4.2 The speed of propagation of waves and dispersion

The phase constant and phase velocity of a plane wave propagating in free space were defined in Section 2.2.4 and extended to lossy dielectrics in (2.50) and (2.51), respectively. In lossless dielectrics, the phase constant (wavenumber), k is linearly dependent on frequency ($k = \omega\sqrt{\mu\varepsilon}$) and as a consequence, the phase velocity is independent of frequency ($v_p = \omega/k = 1/\sqrt{\mu\varepsilon}$) as can be seen from (2.31) and (2.36). On the other hand, in lossy dielectrics, except in cases of very low loss, the phase constant is a rather complex function of frequency and conductivity, leading to a phase velocity that is frequency dependent as can be seen in (2.51). As long as the wave is monochromatic, this simply means that the wave propagates at a different velocity than another wave at a different frequency. However, if the wave is not monochromatic, that is if it is made of, say, the superposition of harmonics, information at each constituent frequency in the bandwidth will propagate at a different frequency. The consequence of this is a distortion of the information carried by the wave. This is referred to as dispersion. This situation is common in a number of applications. In communication, where information requires a relatively wide bandwidth, the frequency dependence of velocity means that lower and higher frequency components arrive at the receiver at different times leading to distortions and loss of information. In GPR applications where pulsed operation is common, this phenomenon is significant because of the wideband generated by very short pulses. The velocity of electromagnetic waves under these conditions is modified through the concept of group velocity. In GPR, the related group delay due to the fact that group velocity is different than phase velocity is often used. This will be defined in the following chapter, whereas the source of this delay is discussed here in terms of group velocity.

2.4.3 Group velocity

Group velocity is the velocity of a wave packet consisting of a narrow range or band of frequencies. An example akin to this is an FM wave as used in FM radio transmission. In this type of wave, a carrier wave at an angular frequency ω_0 is modulated by another wave of angular frequency $\Delta\omega \ll \omega_0$. The angular frequency of the wave will vary between $\omega_0 - \Delta\omega$ and $\omega_0 + \Delta\omega$. Clearly, one cannot now talk about the phase velocity of the wave because phase velocity is only defined for a single frequency. To define the group velocity of a packet of waves, it is useful to consider the case of amplitude modulation. Consider two waves with the same amplitude and propagating in the same direction, but the two waves are at slightly different frequencies. One wave is at an angular frequency $\omega_1 = \omega_0 - \Delta\omega$ and the other at $\omega_2 = \omega_0 + \Delta\omega$. The amplitudes of the waves are E (for simplicity) and the waves propagate in the z direction in a lossless medium. The phase constants of each of the waves are written from the definition $\beta = \omega\sqrt{\mu\varepsilon}$ [rad/m]. Therefore, $\beta_1 = \beta_0 - \Delta\beta$ and $\beta_2 = \beta_0 + \Delta\beta$. With these, the waves are

$$E_1(z,t) = E\cos(\omega_1 t - \beta_1 z) \quad [\text{V/m}] \tag{2.60}$$

$$E_2(z,t) = E\cos(\omega_2 t - \beta_2 z) \quad [\text{V/m}] \tag{2.61}$$

The sum of these two waves gives the total wave:

$$\begin{aligned} E(z,t) &= E_1(z,t) + E_2(z,t) \\ &= [2E\cos(\Delta\omega t - \Delta\beta z)]\cos(\omega_0 t - \beta_0 z) \quad [\text{V/m}] \end{aligned} \tag{2.62}$$

This is a wave with amplitude equal to the sum of the amplitudes of the individual waves and a fundamental or carrier frequency ω_0. The amplitude of the wave varies cosinusoidally with frequency $\Delta\omega$ as can be seen in Figure 2.8. The carrier travels at a velocity v_p, which is calculated (see (2.36)) as for any monochromatic waves. By assuming a constant point on the carrier, the phase velocity of the single frequency carrier is

$$\begin{aligned} \omega_0 t - \beta_0 z &= const. \rightarrow \frac{d(\omega_0 t - \beta_0 z)}{dt} \\ &= \omega_0 - \beta_0\frac{dz}{dt} = 0 \quad \rightarrow \quad v_p = \frac{dz}{dt} = \frac{\omega_0}{\beta_0} \quad [\text{m/s}] \end{aligned} \tag{2.63}$$

This is as expected for a monochromatic plane wave. The phase velocity of the modulating signal may be written from (2.62) as the group velocity v_g:

$$\begin{aligned} \Delta\omega t - \Delta\beta z &= const. \rightarrow \frac{d(\Delta\omega t - \Delta\beta z)}{dt} \\ &= \Delta\omega - \Delta\beta\frac{dz}{dt} = 0 \quad \rightarrow \quad v_g = \frac{dz}{dt} = \frac{\Delta\omega}{\Delta\beta} \quad [\text{m/s}] \end{aligned} \tag{2.64}$$

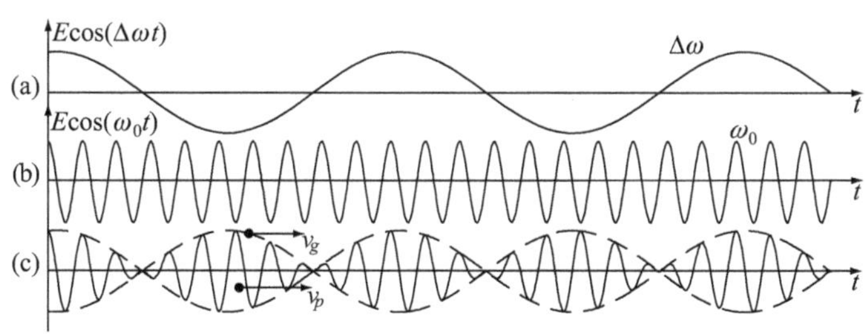

Figure 2.8 Amplitude modulation: (a) the modulating signal, (b) the high-frequency carrier, and (c) the amplitude-modulated carrier. Signals shown at z =0

The group velocity (in this case the velocity of the modulation) is defined as a limit:

$$v_g = \lim_{\Delta\beta \to 0} \frac{1}{\Delta\beta/\Delta\omega} = \frac{1}{d\beta/d\omega} \quad \text{[m/s]} \tag{2.65}$$

This velocity is called the group velocity or the velocity of the wave packet with a narrow frequency width ($\Delta\omega \ll \omega_0$). The latter is actually an informative name since the modulation, or the group, is traveling at this velocity, which can be very different than the phase velocity. The definition given here does not apply to wide-band signals.

2.4.4 Dispersion

The fact that waves at different frequencies traveling through lossy media propagate at different velocities arrive at the target at different times and with different phase delays leads to distortion of signals received, an effect called dispersion. Although dispersion in very low loss dielectrics such as air is often not a noticeable effect, propagation of waves in soils and structures, which have significant losses, or propagation to long distances in lower loss dielectrics, is almost always dispersive. Dispersion can also occur in transmission lines and waveguides, especially if these are long. A dispersion relation is the relation between β and ω shown in Figure 2.9. Figure 2.9(a) shows a number of nondispersive materials. The relation between β and ω is linear. Therefore, when taking the derivative in (2.65), the result is a straight line whose slope is the phase velocity in the medium. Figure 2.9(b) shows a nonlinear dispersion relation. The group velocity is the tangent to the line describing the ω–β relation at any point on the line, indicating that the group velocity is frequency dependent. The line asymptotically tangential to the curve is the lossless phase velocity (since, as frequency approaches infinity, $\sigma/\omega\varepsilon$ approaches zero and the material becomes lossless).

In most dispersive materials, the phase velocity decreases as frequency increases. This dispersion is called normal dispersion. In some other materials, the phase velocity increases with frequency. This is called anomalous dispersion.

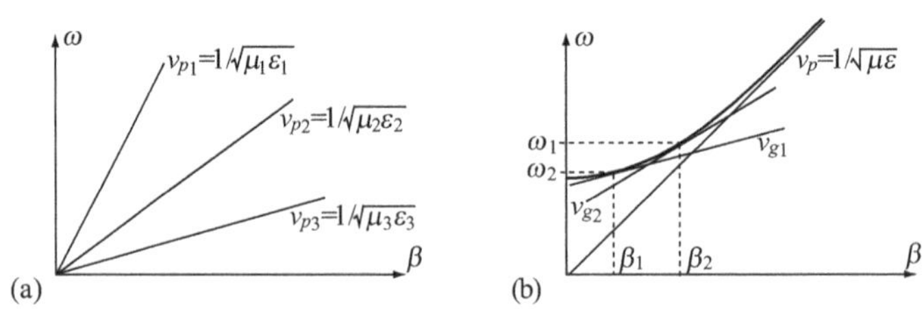

Figure 2.9 Nondispersive and dispersive media: (a) a nondispersive medium has a linear relation between frequency and phase constant and (b) a dispersive medium has a nonlinear relation between frequency and phase constant

2.4.5 Material properties

As was indicated in Chapter 1, Section 1.3.2, properties of materials are central to the successful operation of GPR. The proper identification of materials and their properties plays an important role in the analysis of signals obtained from GPR assessments, and material property models are essential to any attempt at calculation and simulation of assessments. In almost all practical cases, the permeability of materials may be assumed to be constant within the frequency ranges of interest in GPR. Although targets encountered by GPR signals may in fact be ferromagnetic, the signals reflected from the targets are a function of the target's wave impedance and that is more dependent on the complex permittivity (see (2.53)) than on permeability. As well, in most practical cases, permeability of ferromagnetic materials at high frequencies is rather low. Because of that, the properties of the host material (soil, concrete, etc.) are more important for the ability of GPR to detect targets because of the effects of attenuation, dispersion as well as polarization of charges within the medium. For practical purposes, conductivity is viewed as a constant, independent of frequency, whereas polarization is typically frequency dependent. However, the permittivity of media is related to the polarization of charges in response to the electric field intensity in lossless or lossy dielectrics and hence is also frequency dependent. Complex permittivity is a property of lossy dielectrics and also depends on conductivity (see (2.27)). For analysis and in particular for simulation and numerical calculations, the models that are used to describe material properties are of paramount importance. In particular, models for polarization are important for the proper modeling of the complex permittivity of the medium—either lossy or lossless. There are a number of models for permittivity that may be used to define permittivity and are useful in defining the dispersive nature of media.

In general, all dispersive materials found as hosts in GPR application can be modeled using one of the available models for polarization. The most common are the Debye and Lorentz representations [11]. The Lorentz representation is used for most of the solid materials (ionic and electronic polarization) and is based on the system resonant frequencies originating in the displacement of bound charges. The Debye model is a relaxation model used for polar liquids (dipole or orientational polarization), whereby the rate of alignment of the molecules is predicted and the time needed to orient the polarization or relaxation time is defined. There is also a simpler model called the static polarization model that, because it is based on simplistic static considerations, is not appropriate for use in the GPR context but affords a simple, starting point in understanding the idea of polarization. The static polarization model is introduced next, followed by the Debye and Lorenz models.

2.4.5.1 Static polarization and the concept of relative permittivity

In the simplest form, one can view polarization as the effect of forces on bound charges in the atom. In the presence of an external electric field intensity and because charges are bound in atoms, they are pulled toward the opposite polarity to form a local dipole as shown in Figure 2.10. This alignment then produces local, internal fields in directions opposite the external field which, in effect, reduce the total electric

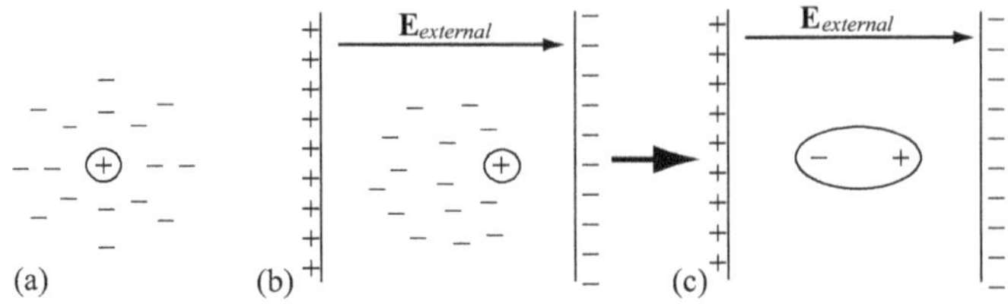

Figure 2.10 Polarization of charges in a dielectric: (a) a charge-neutral atom, (b) the electron cloud is shifted slightly due to the external electric field, and (c) the net effect is an equivalent dipole

field intensity within the dielectric [1,2,11]. The polarization vector may be written formally as the vector sum of the polarizations of individual atoms:

$$\mathbf{P} = \lim_{\Delta v \to 0} \frac{1}{\Delta v} \sum_{i=1}^{N} \mathbf{p}_i \quad \left[\mathrm{C/m^2}\right] \tag{2.66}$$

Although this formula does not allow calculation of polarization for obvious reasons, it does show dependence on the number of dipoles and their individual strength. The polarization may be uniform or nonuniform. Given an external electric field intensity **E** in the absence of the dielectric, the electric flux density is $\varepsilon_0\mathbf{E}$ where ε_0 is the permittivity of free space. In the dielectric, in the presence of a polarization vector **P**, and because of the continuity of the electric flux density across the interface between the dielectric and vacuum, one can write

$$\mathbf{D} = \varepsilon_0\mathbf{E} + \mathbf{P} \quad \left[\mathrm{C/m^2}\right] \tag{2.67}$$

The polarization vector itself is dependent on the external field, since in its absence there cannot be polarization (a residual polarization can exist either temporarily until charges relax to their nonpolarized state or permanently as in electrets). This dependency may be written as

$$\mathbf{P} = \varepsilon_0\chi_e\mathbf{E} \quad \left[\mathrm{C/m^2}\right] \tag{2.68}$$

where χ_e is called the electric susceptibility of the medium and indicates how easy or difficult it is to polarize charges within it. χ_e is a property of the medium and is a dimensionless quantity. This transforms (2.67) into the following:

$$\mathbf{D} = \varepsilon_0\mathbf{E} + \mathbf{P} = \varepsilon_0\mathbf{E} + \varepsilon_0\chi_e\mathbf{E} = \varepsilon_0(1+\chi_e)\,\mathbf{E} \quad \left[\mathrm{C/m^2}\right] \tag{2.69}$$

The quantity $1 + \chi_e$ is a dimensionless quantity called the relative permittivity of the medium.

$$\varepsilon_r = (1 + \chi_e) \quad [\text{dimensionless}] \tag{2.70}$$

Hence, the permittivity of the medium may be written as

$$\varepsilon = \varepsilon_0\varepsilon_r \quad [\mathrm{F/m}] \tag{2.71}$$

The preceding arguments are valid only in the static case. For example, the model does not take into account the possibility of charges relaxing or polarizing in the opposite direction under AC conditions. At low frequencies, one can easily envision the charges following the forces produced by the electric field intensity, since the motion that this would entail is slow. At high frequencies, this cannot happen and therefore this model is not appropriate for any reasonably high frequency. This is borne by experience; the permittivity of media typically goes down as frequency increases, whereas the model in (2.71) is constant. For these reasons, the model described here can only be used for static permittivity or for low frequencies as an approximation.

2.4.5.2 Debye model for polarization

The Debye model for dipolar or orientational polarization is based on the relaxation response of a noninteracting distribution of dipoles in an external alternating electric field [11–13]. It is given by

$$\tau \frac{\delta \mathbf{P}}{\delta t} + \mathbf{P} = \varepsilon_0 \left(\varepsilon_r - \varepsilon_{r\infty}\right) \mathbf{E} \tag{2.72}$$

where ε_r is the static relative permittivity defined in (2.70), $\varepsilon_{r\infty}$ is the relative permittivity at infinite frequency, and τ is the relaxation time. The classical Debye model for relative permittivity as a function of frequency is

$$\varepsilon_r(\omega) = \varepsilon_{r\infty} + \frac{\Delta\varepsilon}{1 + j\omega\tau} \tag{2.73}$$

where $\Delta\varepsilon = \varepsilon_r - \varepsilon_{r\infty}$ is the change in relative permittivity and τ is the characteristic relaxation time relaxation time [11].

In the more general case, the medium may have multiple characteristic relaxation times (as indicated by resonant frequencies or poles), the relative permittivity becomes

$$\varepsilon_r(\omega) = \varepsilon_{r\infty} + \sum_{k=1}^{K} \frac{\Delta\varepsilon_k}{1 + j\omega\tau_k} \tag{2.74}$$

where now $\Delta\varepsilon_k = \varepsilon_{rk} - \varepsilon_{r\infty}$ is the change in relative permittivity due to the kth Debye pole and τ_k is the pole relaxation time.

The general model for complex permittivity may be written from (2.27) by simply replacing ε_r with $\varepsilon_{r\infty}$ from (2.73) or (2.74):

$$\varepsilon_c(\omega) = \varepsilon_0 \left(\varepsilon_{r\infty} + \sum_{k=1}^{K} \frac{\Delta\varepsilon_k}{1 + j\omega\varepsilon} - j\frac{\sigma}{\omega\varepsilon_0} \right) \quad \text{[F/m]} \tag{2.75}$$

The Debye model is much more complex than the static model but it allows better modeling of dispersion effects and the correct loss tangent in the medium.

There are many adjustments and modifications of the Debye model to fit specific materials and conditions [11,13–17], but these are not critical for understanding and are not discussed here. They may be used as necessary for specific applications but are not often encountered in the context of GPR.

2.4.5.3 Lorentz model for polarization

The Lorentz model for polarization starts with the physical assumption that electrons behave like simple harmonic oscillators and hence their motion is described by a second-order differential equations of a damped driven oscillator [18–20]:

$$\frac{\delta^2 \mathbf{P}}{\delta t^2} + \frac{1}{\tau}\frac{\delta \mathbf{P}}{\delta t} + \omega_0^2 \mathbf{P} = \varepsilon_0 \Delta\varepsilon_r \omega_0^2 \mathbf{E} \tag{2.76}$$

where $\Delta\varepsilon_r = \varepsilon_r - \varepsilon_{r\infty}$ and ω_0 is the resonant frequency of the material corresponding to its relaxation time τ, and the driving function is the electric field intensity on the right-hand side of the relation. The damping factor is $1/2\tau$. The frequency-dependent relative permittivity then becomes

$$\varepsilon_r(\omega) = \varepsilon_{r\infty} + \frac{\Delta\varepsilon_r \omega_0^2}{\omega_0^2 + j2\omega\delta - \omega^2} \tag{2.77}$$

As with the Debye model, if the medium is such that it has multiple relaxation time constants (multiple resonant frequencies), the multipole approximation to the relative permittivity may be written as

$$\varepsilon_r(\omega) = \varepsilon_{r\infty} + \sum_{k=1}^{K} \frac{\Delta\varepsilon_{rk}\omega_k^2}{\omega_k^2 + j2\omega\delta_k - \omega^2} \tag{2.78}$$

where $\Delta\varepsilon_{rk} = \varepsilon_{rk} - \varepsilon_{r\infty}$ is the change in relative permittivity relative to the kth pole, δ_k the damping factor for the kth pole, and ω_k the resonant frequency for the kth pole. With these, the complex permittivity may be written in a form similar to (2.75)

$$\varepsilon_c(\omega) = \varepsilon_0\left(\varepsilon_{r\infty} + \sum_{k=1}^{K} \frac{\Delta\varepsilon_{rk}\omega_k^2}{\omega_k^2 + j2\omega\delta_k - \omega^2} - j\frac{\sigma}{\omega\varepsilon_0}\right) \quad [\text{F/m}] \tag{2.79}$$

As with the Debye model, the Lorentz model has also inspired variations and adaptations. The models for polarization discussed here and others as well as variations on the present models [21–24] are essential in the understanding of wave propagation through media. Many of the models have been initially developed for the treatment of light waves but they apply to lower frequency electromagnetic waves and even to heat radiation [25]. Finally, it should be mentioned that more often than not, many materials encountered in GPR assessments are common media such as soils and concretes and these either have known or estimated properties, including their variation with frequency, and are well documented. In other cases, materials can be measured if need be with any of the methods available for this purpose [26–32]. In cases where media are mixtures of constituents, a number of methods of electromagnetic homogenization may be employed to obtain the properties of the mixture [33–37]. (Electromagnetic homogenization is the process of estimating the effective electromagnetic properties of composite materials in the long-wavelength regime, that is, on the scale of non-homogeneities that are much smaller than the wavelengths involved.)

2.4.6 Homogeneity, linearity, and anisotropy of materials

Now that the electric properties of materials have been defined through permittivity, conductivity, and permeability (and, indirectly, through polarization), it is appropriate to classify these properties. Specifically, there are three important classifiers: linearity, homogeneity, and isotropy of materials. A material is linear in a particular property (such as permittivity) if the property does not change when the fields change. A material of this type is said to be linear even though its properties may vary with frequency as discussed earlier. This property should not be confused with linear dependency. For a medium to be linear in a property, that property must be independent of the applied field. Most practical materials encountered in work with GPR are linear in conductivity and permittivity. Similarly, most materials are linear in permeability, although ferromagnetic materials are likely to possess nonlinear permeability. Again, at high frequencies, these are of reduced importance simply because permeability is typically low and for most materials it is the same as that of free space.

A medium is said to be homogeneous in a property if that property does not vary from point to point in the medium. Uniform media, with constant density, are usually of this type. On the other hand, soils and other materials, where density and humidity can vary and composition of which can vary considerably with depth, are clearly inhomogeneous. The effect that inhomogeneity has on GPR signals depends on the level of inhomogeneity and its nature but as a matter of principle, it changes the propagation properties of the medium (attenuation and phase constants) and hence the signal received. It can affect depth detection and can scatter the fields, reducing sensitivity.

Isotropy of a medium relates to the properties of the medium with respect to direction. If a medium is uniform in all directions, it is said to be isotropic. Anisotropic materials may exhibit marked differences in propagation of waves in different directions within the medium. Media such as crystals are known to be highly anisotropic and anisotropy can have useful applications. However, host materials normally encountered in GPR assessment are almost always isotropic. It should also be mentioned that many of the assumptions used in this chapter and elsewhere are based on isotropy of materials. For example, the constitutive relations for permittivity, permeability, and conductivity in (2.9)–(2.11) were based on single-valued (isotropic) material properties, whereas the fields **E**, **H**, **B**, and **J** are three-dimensional vectors. This assumption also means that, for example, **E** and **D** are in the same direction in space. If permittivity, permeability, or conductivity were anisotropic, the fields they relate would not be colinear and the property itself would have to be described as a tensor.

Materials with linear, homogeneous, and isotropic properties are called simple materials. A medium may be homogeneous and isotropic but nonlinear (iron, nickel, ferrites, etc.) or they may be homogeneous, linear, and anisotropic (wood, layered materials, crystals, some glasses, most composite materials, etc.). Inhomogeneous materials may be isotropic or anisotropic. For example, a layered material such as plywood or a carbon composite may be both anisotropic and inhomogeneous. The atmosphere, because of variations in pressure, humidity, and temperature, is inhomogeneous on the large scale, whereas most soils are inhomogeneous either on the small scale (variations in grain sizes in sand) or on the larger scale due to variations in constituents, moisture, and the like. Some soils, such as clay may be highly homogeneous at almost any scale, whereas others may be homogeneous on the scale used for GPR assessment. Since the scale for these purposes is the wavelength, media in which inhomogeneities are small compared to the wavelength may be considered homogeneous.

It is common in GPR to assume the host material to be a simple material, often for the lack of better information on the medium, whereas in other cases, such as in concrete, the assumption is based on the knowledge of the properties of concrete. Any inhomogeneity that may be detected during assessment is then considered a flaw in concrete rather than a property of the medium.

2.5 Reflection, transmission, refraction, scattering, and diffraction of electromagnetic waves

The previous sections discussed the propagation of electromagnetic waves in what may be called an unbounded medium in that the effects of interfaces between media (i.e., an inclusion or a target in the path of the wave) were not discussed. Allusion was made in (2.32) to forward- and backward-propagating waves, but the source of the backward-propagating wave was not indicated. The forward-propagating wave may be generated by an antenna and will propagate away from the source but why should there be a backward-propagating wave? The reason can only be that the wave, or part of it, is reflected back from a medium, such as a target. In fact, without the possibility of waves being "returned" from targets, the whole idea of GPR is irrelevant. Thus, it is essential to take into account any discontinuity in the background or host medium to see how waves are reflected back and to analyze their properties. But there is more than just reflection. If part of the wave is reflected then part of it must transmit through the medium that causes the reflection. These concepts will be described next. In addition, the concepts of refraction and scattering of electromagnetic waves are discussed briefly since these, in addition to reflection and transmission affect the signals received from targets and therefore their analysis.

2.5.1 Reflection and transmission of electromagnetic waves at a general interface

The discussion starts with a general interface as shown in Figure 2.11, which may represent the interface between air and soil or between soil and a target in a GPR configuration. It can, however, represent any other discontinuity in the path of the electromagnetic wave. Both media are assumed to be lossy with different material properties as indicated. The electric and magnetic field intensities are perpendicular to each other and to the direction of propagation (plane waves), and the incident wave propagates toward the interface at an incidence angle θ_i with the normal to the interface. The reflected wave propagates away from the interface in medium (1) at a

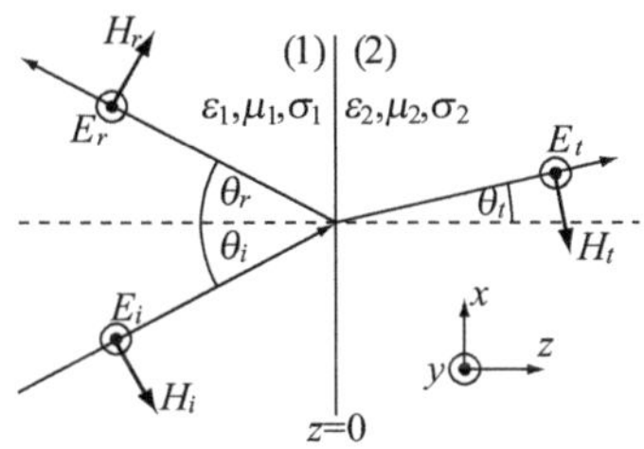

Figure 2.11 Incident, reflected, and transmitted (refracted) waves at a general interface

reflection angle θ_r. A transmitted wave propagates in medium (2) at a transmission angle θ_t to the normal [1,2]. In this configuration, the electric field intensity in the incident wave is perpendicular to the page. The plane of the page, that is, the plane formed by the normal to the interface and the direction of propagation of the incident wave is called the plane of incidence, and if the electric field intensity is perpendicular to this plane, as in the figure, the wave is said to be perpendicularly polarized. An electric field intensity in the plane of incidence is said to have parallel polarization. Each polarization leads to different behavior of the waves at the interface. For the ease of analysis, the interface is placed at $z = 0$ as shown.

To define the reflection and transmission at the interface, the electric and magnetic field intensities on both sides of the interface are written in general terms, and then the interface conditions at the interface are invoked to match the field components at the interface. The incident wave propagates toward the interface and, assuming it travels a distance r_1, the incident electric field intensity may be written as

$$\mathbf{E}_i = \hat{\mathbf{y}}E_i e^{-\alpha_1 r_1} e^{-j\beta_1 r_1} = \hat{\mathbf{y}}E_i e^{-\alpha_1 (z\cos\theta_i + x\sin\theta_i)} e^{-j\beta_1 (z\cos\theta_i + x\sin\theta_i)} \tag{2.80}$$

$$\begin{aligned}\mathbf{H}_i &= \left(-\hat{\mathbf{x}}\frac{E_i}{\eta_1}\cos\theta_i + \hat{\mathbf{z}}\frac{E_i}{\eta_1}\sin\theta_i\right) e^{-\alpha_1 r_1} e^{-j\beta_1 r_1} \\ &= \left(-\hat{\mathbf{x}}\frac{E_i}{\eta_1}\cos\theta_i + \hat{\mathbf{z}}\frac{E_i}{\eta_1}\sin\theta_i\right) e^{-\alpha_1 (z\cos\theta_i + x\sin\theta_i)} e^{-j\beta_1 (z\cos\theta_i + x\sin\theta_i)}\end{aligned} \tag{2.81}$$

where

$$|\mathbf{H}_i| = \frac{|\mathbf{E}_i|}{\eta} \tag{2.82}$$

was used to write the magnetic field intensity. Note also that the direction of propagation of the magnetic field intensity must be toward the interface (as indicated by the Poynting vector), and hence the incident magnetic field intensity must have components in the negative x and positive z directions as shown. The form on the right-hand side of either equation shows that if the distance is $r_1 = z\cos\theta_i + x\sin\theta_i$, the result is identical. However, in this form, the wave is seen as propagating a distance $z\cos\theta_i$

in the z direction and a distance $x \sin\theta_i$ in the x direction rather than a distance r_1 in the direction of propagation of the wave. This form will allow the application of the interface conditions for the tangential components of the electric and magnetic field intensities as

$$E_{1tang} = E_{2tang}, \qquad H_{1tang} = H_{2tang} \tag{2.83}$$

The reflected wave must propagate away from the interface and hence in the positive x and negative z directions. These may be written directly from Figure 2.11 as

$$\mathbf{E}_r = \hat{\mathbf{y}} E_r e^{-\alpha_1(-z\cos\theta_r + x\sin\theta_r)} e^{-j\beta_1(-z\cos\theta_r + x\sin\theta_r)} \tag{2.84}$$

$$\mathbf{H}_r = \left(\hat{\mathbf{x}}\frac{E_r}{\eta_1}\cos\theta_r + \hat{\mathbf{z}}\frac{E_r}{\eta_1}\sin\theta_r\right) e^{-\alpha_1(-z\cos\theta_r + x\sin\theta_r)} e^{-j\beta_1(-z\cos\theta_r + x\sin\theta_r)} \tag{2.85}$$

In medium (2), the wave propagates away from the interface and has the following forms:

$$\mathbf{E}_t = \hat{\mathbf{y}} E_t e^{-\alpha_2(z\cos\theta_t + x\sin\theta_t)} e^{-j\beta_2(z\cos\theta_t + x\sin\theta_t)} \tag{2.86}$$

$$\mathbf{H}_t = \left(-\hat{\mathbf{x}}\frac{E_t}{\eta_2}\cos\theta_t + \hat{\mathbf{z}}\frac{E_t}{\eta_2}\sin\theta_t\right) e^{-\alpha_1(z\cos\theta_t + x\sin\theta_t)} e^{-j\beta_1(z\cos\theta_t + x\sin\theta_t)} \tag{2.87}$$

The wave impedances η_1, η_2, attenuation constants α_1, α_2, and phase constants β_1 and β_2 are known from the material properties (see (2.49), (2.50), and (2.53)). Although $\mathbf{E}_i$ is given since the incident wave is the source of the wave, $\mathbf{E}_r$ and $\mathbf{E}_t$ are unknown and must be found from the interface conditions. These are applied at $z = 0$ on the tangential components (the x or y components) of the electric and magnetic field intensities. These are as follows:

$$\mathbf{E}_{it} = \hat{\mathbf{y}} E_i e^{-\alpha_1 x\sin\theta_i} e^{-j\beta_1 x\sin\theta_i} \tag{2.88}$$

$$\mathbf{H}_{it} = -\hat{\mathbf{x}}\frac{E_i}{\eta_1}\cos\theta_i e^{-\alpha_1 x\sin\theta_i} e^{-j\beta_1 x\sin\theta_i} \tag{2.89}$$

$$\mathbf{E}_{rt} = \hat{\mathbf{y}} E_r e^{-\alpha_1 x\sin\theta_r} e^{-j\beta_1 x\sin\theta_r} \tag{2.90}$$

$$\mathbf{H}_{rt} = \hat{\mathbf{x}}\frac{E_r}{\eta_1}\cos\theta_i e^{-\alpha_1 x\sin\theta_r} e^{-j\beta_1 x\sin\theta_r} \tag{2.91}$$

$$\mathbf{E}_{tt} = \hat{\mathbf{y}} E_t e^{-\alpha_2 x\sin\theta_t} e^{-j\beta_2 x\sin\theta_t} \tag{2.92}$$

$$\mathbf{H}_{tt} = -\hat{\mathbf{x}}\frac{E_t}{\eta_2}\cos\theta_t e^{-\alpha_1 x\sin\theta_t} e^{-j\beta_2 x\sin\theta_t} \tag{2.93}$$

The total tangential fields in medium (1) and (2) are

$$\mathbf{E}_{1t} = \mathbf{E}_{it} + \mathbf{E}_{rt} = \hat{\mathbf{y}}\left(E_i e^{-\alpha_1 x\sin\theta_i} e^{-j\beta_1 x\sin\theta_i} + E_r e^{-\alpha_1 x\sin\theta_r} e^{-j\beta_1 x\sin\theta_r}\right) \tag{2.94}$$

$$\begin{aligned}\mathbf{H}_{1t} &= \mathbf{H}_{it} + \mathbf{H}_{rt} \\ &= \hat{\mathbf{x}}\left(-\frac{E_i}{\eta_1}\cos\theta_i e^{-\alpha_1 x\sin\theta_i}e^{-j\beta_1 x\sin\theta_i}\right. \\ &\qquad \left.+\frac{E_r}{\eta_1}\cos\theta_r e^{-\alpha_1 x\sin\theta_r}e^{-j\beta_1 x\sin\theta_r}\right)\end{aligned} \tag{2.95}$$

$$\mathbf{E}_{2t} = \hat{\mathbf{y}}E_t e^{-\alpha_2 x\sin\theta_t}e^{-j\beta_2 x\sin\theta_t} \tag{2.96}$$

$$\mathbf{H}_{2t} = -\hat{\mathbf{x}}\frac{E_t}{\eta_2}\cos\theta_t e^{-\alpha_1 x\sin\theta_2}e^{-j\beta_2 x\sin\theta_t} \tag{2.97}$$

Applying now the interface conditions for the electric and magnetic field intensities in (2.83):

$$\begin{aligned}&E_i e^{-\alpha_1 x\sin\theta_i}e^{-j\beta_1 x\sin\theta_i} + E_r e^{-\alpha_1 x\sin\theta_r}e^{-j\beta_1 x\sin\theta_r} \\ &\quad = E_t e^{-\alpha_2 x\sin\theta_t}e^{-j\beta_2 x\sin\theta_t}\end{aligned} \tag{2.98}$$

$$\begin{aligned}&-\tfrac{E_i}{\eta_1}\cos\theta_i e^{-\alpha_1 x\sin\theta_i}e^{-j\beta_1 x\sin\theta_i} + \tfrac{E_r}{\eta_1}\cos\theta_r e^{-\alpha_1 x\sin\theta_r}e^{-j\beta_1 x\sin\theta_r} \\ &\quad = -\tfrac{E_t}{\eta_2}\cos\theta_t e^{-\alpha_1 x\sin\theta_t}e^{-j\beta_2 x\sin\theta_t}\end{aligned} \tag{2.99}$$

To simplify these, note the $E_i e^{-\alpha_1 x\sin\theta_i}$, $E_r e^{-\alpha_1 x\sin\theta_r}$, and $E_t e^{-\alpha_2 x\sin\theta_t}$ are the amplitudes of the incident, reflected, and transmitted waves at the interface. These are denoted simply as E_{i0}, E_{r0}, and E_{t0}. Similarly, the first of Snell's laws of reflection states that the angle of reflection must be equal to the angle of incidence. Substituting these into (2.98) and (2.99) gives

$$(E_{i0} + E_{r0})\,e^{-j\beta_1 x\sin\theta_i} = E_{t0}e^{-j\beta_2 x\sin\theta_t} \tag{2.100}$$

$$\left(-\frac{E_{i0}}{\eta_1}\cos\theta_i + \frac{E_{r0}}{\eta_1}\cos\theta_i\right)e^{-j\beta_1 x\sin\theta_i} = -\frac{E_{t0}}{\eta_2}\cos\theta_t e^{-j\beta_2 x\sin\theta_t} \tag{2.101}$$

for the two sides of (2.100) to be equal, the amplitudes and the phases must satisfy the following:

$$E_{i0} + E_{r0} = E_{t0} \tag{2.102}$$

$$\begin{aligned}\beta_1\sin\theta_i &= \beta_2\sin\theta_t \rightarrow \sin\theta_t = \frac{\beta_1}{\beta_2} \\ \sin\theta_i &= \frac{\omega\sqrt{\mu_1\varepsilon_1}}{\omega\sqrt{\mu_2\varepsilon_2}}\sin\theta_i = \sqrt{\frac{n_1}{n_2}}\sin\theta_i\end{aligned} \tag{2.103}$$

where n_1 and n_2 are the indices of refraction of materials (1) and (2) at the frequency of the wave. The angle of refraction θ_t (also called the transmission angle) is

$$\theta_t = \sin^{-1}\left(\sqrt{\frac{n_1}{n_2}}\sin\theta_i\right) \tag{2.104}$$

Given the relation in (2.104), the relation in (2.101) becomes

$$\left(\frac{E_{i0}}{\eta_1} - \frac{E_{r0}}{\eta_1}\right)\cos\theta_i = \frac{E_{t0}}{\eta_2}\cos\theta_t \tag{2.105}$$

Now that θ_t and θ_r are known in terms of θ_i, the remaining unknowns are E_{r0} and E_{t0}. These are found by solving (2.102) and (2.105). The results are:

$$E_{r0} = \left(\frac{\eta_2\cos\theta_i - \eta_1\cos\theta_t}{\eta_2\cos\theta_i + \eta_1\cos\theta_t}\right)E_{i0} \tag{2.106}$$

$$E_{t0} = \left(\frac{2\eta_2\cos\theta_i}{\eta_2\cos\theta_i + \eta_1\cos\theta_t}\right)E_{i0} \tag{2.107}$$

That is, the reflected and transmitted electric field intensities, and therefore the magnetic field intensities can be calculated from the incident electric field intensity and its angle of incidence and the given wave impedances in the two media. The ratio between the reflected and the incident electric field intensities is known as the reflection coefficient, whereas that between the transmitted and incident electric field intensities is known as the transmission coefficient:

$$\Gamma_\perp = \frac{E_{r1}}{E_{i1}} = \frac{\eta_2\cos\theta_i - \eta_1\cos\theta_t}{\eta_2\cos\theta_i + \eta_1\cos\theta_t} \quad \text{[dimensionless]} \tag{2.108}$$

$$T_\perp = \frac{E_{t2}}{E_{i1}} = \frac{2\eta_2\cos\theta_i}{\eta_2\cos\theta_i + \eta_1\cos\theta_t} \quad \text{[dimensionless]} \tag{2.109}$$

The notation $\perp$ indicates that these are defined for the case of perpendicular polarization of the electric field intensity (Figure 2.11).

One can verify by direct calculation that the following holds:

$$1 + \Gamma_\perp = T_\perp \tag{2.110}$$

With these relations, the reflected, transmitted, and total field in both media can be described accurately. Specifically, the reflected wave that is, in fact, responsible for detection and evaluation of targets in GPR is completely and accurately specified on the basis of the properties of the host and target media and angle of incidence on the target. This may not be as simple as that because the target is not a plane surface and the angle of incidence varies from point to point on the target. For this reason, in GPR it is more appropriate to talk about scattering rather than simple reflection. Scattering is discussed in the following section.

If the incident wave is a parallel polarized wave, that is, that the electric field intensity in Figure 2.11 is parallel to the plane of incidence, and following a similar process, the reflection and transmission coefficients for parallel polarization may be written as [1,2]

$$\Gamma_\parallel = \frac{\eta_2\cos\theta_t - \eta_1\cos\theta_i}{\eta_2\cos\theta_t + \eta_1\cos\theta_i} \quad \text{[dimensionless]} \tag{2.111}$$

$$T_\parallel = \frac{2\eta_2\cos\theta_i}{\eta_2\cos\theta_t + \eta_1\cos\theta_i} \quad \text{[dimensionless]} \tag{2.112}$$

and, unlike perpendicular polarization

$$1 + \Gamma_{||} = T_{||}\left(\frac{\cos\theta_t}{\cos\theta_i}\right) \tag{2.113}$$

If the angle of incidence happens to be zero (i.e., the wave is incident perpendicular to the surface), the reflection and transmission coefficients for either parallel or perpendicular polarization become

$$\Gamma = \frac{\eta_2 - \eta_1}{\eta_2 + \eta_1} \quad \text{[dimensionless]} \tag{2.114}$$

$$T = \frac{2\eta_2}{\eta_2 + \eta_1} \quad \text{[dimensionless]} \tag{2.115}$$

This can be understood from the fact that for normal incidence, all fields are tangential to the surface and there cannot be any distinction between perpendicular and parallel polarization. Also, in this case $1 + \Gamma = T$. These relations indicate that the larger the difference in the wave impedance between target and host medium, the larger the reflection coefficient. They also show that the reflection and transmission coefficients are in general complex because the wave impedance may be (and often is) complex as can be seen in (2.53).

2.5.2 Refraction, diffraction, and scattering of electromagnetic waves

The concept of refraction was introduced briefly in the previous subsection, that is, the change in direction of the electromagnetic wave as it propagates across the interface between two different media. Refraction affects the transmitted wave as, for example, from air into the host medium or the reflected (or scattered) wave transmitted from the host medium into air and therefore onto the receiving antenna. The refraction angle was obtained in (2.104) in terms of the indices of refraction of the media n_1 and n_2. Since the index of refraction of a medium is $n = \sqrt{\varepsilon_r\mu_r}$, the angle of refraction may be written as

$$\theta_t = \sin^{-1}\left(\sqrt{\frac{\varepsilon_{r1}\mu_{r1}}{\varepsilon_{r2}\mu_{r2}}}\sin\theta_i\right) \tag{2.116}$$

Clearly, the refraction angle for a wave propagating from one medium (1) into a second (2) is larger than the incident angle, if the product of relative permittivity and relative permeability of medium (1) is larger than medium (2). This may lead to a situation called total reflection without transmission into medium (2). The main effect of refraction that is of real consequence is on the footprint of the radar beam inside the host medium. For a wave in GPR propagating from air into any medium, the refraction angle is lower than the incident angle, leading to a narrowing of the beam and reduction in the footprint in comparison to the equivalent footprint in air at the same distance from the antenna. This will be discussed in the following chapter but was also alluded to in Chapter 1.

Diffraction of electromagnetic waves is the bending of waves around the edges of an obstruction or through an aperture. It is a very common and important aspect of propagation of light waves, but it can occur at any frequency to some degree. Clear diffraction of sound waves and ocean waves can also be observed. As with all propagation properties of waves, it is tied to wavelength and no observable diffraction occurs when the obstruction is large compared to the wavelength. In GPR, diffraction is often used to image surfaces and to reduce the effects of surface effects on the image. Nevertheless, this subject will not be pursued here as it properly belongs in the realm of signal and image processing.

Scattering is a process by which an incoming electromagnetic wave encountering an object or an assembly of objects such as particles, a distribution of objects, and similar conditions (the scatterer or scattering system) loses energy that is absorbed by the scatterer and which is then either entirely or partially redirected or reradiated by the scatterer in a portion of space or in the whole of space. The scattering can be uniform or nonuniform and may be either at the original frequency of the electromagnetic wave or at a different frequency. For the purpose of this discussion, one can think of scattering as a form of more generalized reflection, especially if one concentrates on the observable effects rather than the mathematical details of scattering. There are a number of mechanisms that define scattering. If the size of the scatterers is small compared to wavelength, the process is called Rayleigh scattering [38–40], viewed as elastic scattering, meaning that the transfer of energy to the scatterers is minimal. Waves scattered in soils and structures follow this model especially in the longer wavelengths used in GPR for deep penetration in media. In many cases, scattering in host media can be modeled by Rayleigh scattering. In larger scatterers, such as targets in GPR, which may be of the same order of magnitude as the wavelength, a more appropriate model is Mie scattering [6,41–43]. In both models, the basic theory deals with spheroidal scatterers, although extensions to other shapes do exist. Delving into the theory either Rayleigh or Mie scattering is beyond the scope of this chapter, but there exists a simple model that is often used in radar to define radar or scattering cross-section [1,44–47] or to estimate the range of radar for a given target with known scattering cross-section. This model is discussed here as it is useful in scattering from objects large compared to the wavelength.

In radar, including GPR, the amount of scattered power by an object (scatterer) depends on a number of properties, including size, shape, and composition of the scatterer. Now, suppose that the transmitting antenna transmits a power P_{rad}. Part of this power may be lost through attenuation and other interactions such as reflections off the surface. A small part of the power intercepts the scatterer is scattered and propagates back to the antenna. Only a small part of the power transmitted reaches back to the receiving antenna. One of the most basic of radar properties that affects the power received back from the scatterer is the radar or scattering cross-section. To define the scattering cross-section, consider Figure 2.12. In Figure 2.12(a), the transmitting antenna emits power P_{rad}. How much of this power is reflected by the scatterer? To calculate that power, the equivalent area of the scatterer is assumed to be σ (no relation to conductivity which uses the same symbol). This equivalent area is the scattering cross-section of the target. If the time-averaged power density at the location of the scatterer is $\mathscr{P}_i$, the total power scattered by the target is $\sigma \mathscr{P}_i$ (Figure 2.12(b)). This scattered power can be viewed as a radiated power, transmitted

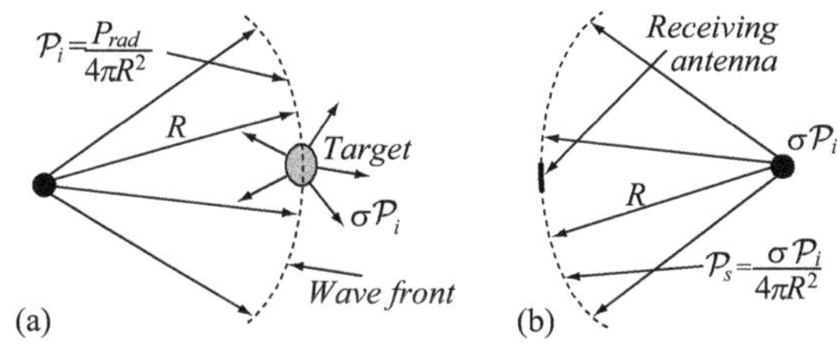

Figure 2.12 (a) Power at the target and (b) power scattered by the target, part of which reaches back at the receiving antenna

by the scatterer. This power produces a scattered power density at the receiving antenna, which may be written (assuming uniform distribution of the scattered field since the scattering cannot be, in general, assumed to be directive) as [1,44]

$$\mathscr{P}_s = \frac{\sigma \mathscr{P}_i}{4\pi R^2} \quad \left[\mathrm{W/m^2}\right] \tag{2.117}$$

The scattering cross-section is therefore

$$\sigma = 4\pi R^2 \frac{\mathscr{P}_s}{\mathscr{P}_i} \quad \left[\mathrm{m^2}\right] \tag{2.118}$$

The scattering cross-section is defined as the ratio of the scattered power by the target divided by the time-averaged power density at the location of the target. However, the incident time-averaged power density at the target depends on the radiated power as

$$\mathscr{P}_i = \frac{P_{rad}}{4\pi R^2} D_t(\theta, \phi) \quad \left[\mathrm{W/m^2}\right] \tag{2.119}$$

where $D_t(\theta, \phi)$ is the directivity of the transmitting antenna, and θ, ϕ are the elevation and azimuthal angles in a spherical system of coordinates centered at the antenna (directivity of antennas will be discussed in the following chapter). Substituting in (2.117), the scattered power density at the receiving antenna is

$$\mathscr{P}_s = \frac{\sigma P_{rad} D_t(\theta, \phi)}{\left(4\pi R^2\right)^2} \quad \left[\mathrm{W/m^2}\right] \tag{2.120}$$

The total power received by the antenna depends on the scattered power density and the effective area of the receiving antenna A_{er}. Thus, the received power is

$$P_r = A_{er} \mathscr{P}_s = \frac{A_{er} \sigma P_{rad} D_t(\theta, \phi)}{\left(4\pi R^2\right)^2} \quad [\mathrm{W}] \tag{2.121}$$

Now, using the definition of effective area of the receiving antenna as

$$A_{er} = D_r(\theta, \phi) \frac{\lambda^2}{4\pi} \quad \left[\mathrm{m^2}\right] \tag{2.122}$$

where $D_r(q,f)$ is the directivity of the receiving antenna. Substituting this in (2.121)

$$P_r = \frac{\sigma P_{rad} D_t(\theta,\phi) D_r(\theta,\phi) \lambda^2}{(4\pi)^3 R^4} \quad [\mathrm{W}] \tag{2.123}$$

Dividing by P_{rad} gives the ratio of received to radiated power as

$$\frac{P_r}{P_{rad}} = \frac{\sigma D_t(\theta,\phi) D_r(\theta,\phi) \lambda^2}{(4\pi)^3 R^4} \tag{2.124}$$

In the particular case in which the transmitting antenna is used for receiving as well (A-static radar applications, pulsed radar), $D_r(\theta,\phi) = D_t(\theta,\phi) = D(\theta,\phi)$:

$$\frac{P_r}{P_{rad}} = \frac{\sigma \lambda^2}{(4\pi)^3 R^4} D^2(\theta,\phi) = \frac{\sigma A_e^2}{4\pi R^4 \lambda^2} \tag{2.125}$$

The radar cross-section can then be written as

$$\sigma = \frac{P_r 4\pi R^4 \lambda^2}{P_{rad} A_e^2} \quad [\mathrm{m}^2] \tag{2.126}$$

Equation (2.125) is called the radar equation. It was developed here in terms of directivity and effective area. Because directivity depends on direction in space (see (2.122) and (2.125)), the radar equation and the radar cross-section are also dependent on direction in space. An alternative form of the equations recasts the effective area in terms of maximum directivity resulting in maximum effective area and maximum range in the radar equation. By replacing $D_r(\theta,\phi) = D_{\max}$ and $D_t(\theta,\phi) = D_{\max}$, (2.122) becomes

$$A_{er\text{-}\max} = D_{\max} \lambda^2 4\pi \quad [\mathrm{m}^2] \tag{2.127}$$

As a consequence, (2.125) becomes

$$\frac{P_r}{P_{rad}} = \frac{\sigma \lambda^2 D_{\max}^2}{(4\pi)^3 R_{\max}^4} \tag{2.128}$$

Recalling that the gain is equal to the maximum directivity for lossless antennas and, that the gain for practical antennas takes into account the losses in the antennas as will be detailed in the next chapter, it is also common to write the radar equation in terms of the gain:

$$\frac{P_r}{P_{rad}} = \frac{\sigma \lambda^2 G^2}{(4\pi)^3 R_{\max}^4} \tag{2.129}$$

Any of the forms in (2.125), (2.128), or (2.129) is an appropriate form for the radar equation. It was defined in the absence of attenuation and any other effects that may occur in the path between antenna and target, and in this sense it can only be viewed as a best case estimate. On the other hand, scattering by the target is assumed to be uniform, whereas in reality, more power may be directed back to the antenna than, for example, in directions away from the antenna, depending on size and shape of the target. In spite of these assumptions, the model is useful in initial estimation either of signal strength [Equation (2.125) or (2.129)], or, alternatively, of radar cross-section

[Equation (2.126)]. The effects of attenuation and other losses are often incorporated in the radar equation to improve the model but at this juncture, the simple lossless model based on uniform scattering is sufficient to give an idea of the primary issues and to point to improvements that can be made to the model.

2.6 Summary

It is not possible to cover all intricacies of electromagnetic wave propagation in a short chapter or, indeed in a whole book, especially taking into account the environment in which GPR operates. Nevertheless, the primary concepts and tools needed for an understanding of wave behavior in complex media have been addressed. The user may have to resort to additional material as indicated in the introduction. It is worth mentioning again that most of the discussion was in terms of plane waves and in the frequency-domain. These waves are unique in many ways not the least in that true plane waves do not exist. These assumptions lead to the definition of the important quantities without diminishing their generality. The concepts of complex permittivity, attenuation, phase constant, phase velocity, wave impedance, skin depth, dispersion, polarization as well as reflection, transmission, and scattering are general and their definition through the medium of plane waves is rather straightforward.

References

[1] Ida N. Engineering Electromagnetics, 4th ed. Springer, Heidelberg; 2020.
[2] Cheng DK. Field and Wave Electromagnetics, 2nd ed. Addison-Wesley, Reading, MA; 1992.
[3] Miner GF. Lines and Electromagnetic Fields for Engineers. Oxford University Press, New York, NY; 1996.
[4] Karmel PR, Colef GD, and Camisa RL. Introduction to Electromagnetic and Microwave Engineering. John Wiley and Sons, Inc., New York, NY; 1998.
[5] Balanis CA. Advanced Engineering Electromagnetics, 2nd ed. Wiley, Hoboken, NJ; 2012.
[6] Stratton JA. Electromagnetic Theory. McGraw-Hill, New York, NY; 1941.
[7] Van Bladel J. Electromagnetic Fields. Hemisphere Publishing, New York, NY; 1985.
[8] Harrington RF. Time-Harmonic Electromagnetic Fields. McGraw-Hill, New York, NY; 1961.
[9] Collin RE. Foundations for Microwave Engineering, 2nd ed. McGraw-Hill, New York, NY; 1992.
[10] Pozar DM. Microwave Engineering, 4th ed., Cambridge University Press, Cambridge; 2017.
[11] Jackson JD. Classical Electrodynamics, 3rd ed. John Wiley & Sons, Inc., Hoboken, NJ; 1999.

[12] Griffiths DJ. Introduction to Electrodynamics. Prentice-Hall, Cambridge; 1999.

[13] Scaife B. Principles of Dielectrics (Monographs on the Physics & Chemistry of Materials), 2nd ed. Oxford University Press, Oxford; 1998.

[14] Cole KS and Cole RH. Dispersion and absorption in dielectrics: I – Alternating current characteristics. The Journal of Chemical Physics, 9:(4);1941:341–351.

[15] Havriliak S and Negami S. A complex plane representation of dielectric and mechanical relaxation processes in some polymers. Journal of Chemical Physics, 9 (4);1967:341–351.

[16] Curie J. Recherches sur le pouvoir inducteur spécifique et sur la conductibilité des corps cristallis's. Annales de Chimie et de Physique, 17;1969:384–434.

[17] Schweidler ER. Studien über die Anomalien im Verhalten der Dielektrika (Studies on the anomalous behaviour of dielectrics). Annalen der Physik, 329 (14);1907:711–770.

[18] Lorentz HA. The Theory of Electrons and Its Applications to the Phenomena of Light and Radiant Heat. Dover Books, New York, NY; 2011.

[19] Feynman RP, Leighton RB, and Sands M. The Feynman Lectures on Physics, Vol. 2, Ch. 32. Addison-Wesley, Reading, MA; 1964.

[20] Oughstun KE and Cartwright NA. On the Lorentz-Lorenz formula and the Lorentz model of dielectric dispersion. Optics Express, 11 (13);2003: 1541–1546.

[21] Drude P. Zur Elektronentheorie der Metalle. Annalen der Physik, 30r (3);1900:566–813.

[22] Dressel M and Scheffler M. Verifying the Drude response. Annalen der Physik, 15 (7–8);2006:535–544.

[23] Burke PJ, Spielman IB, Eisenstein JP, Pfeiffer LN, and West KW. High frequency conductivity of the high-mobility two-dimensional electron gas. Applied Physics Letters, 76 (6);2000:745–747.

[24] Scheffle M, Jourdan M, and Adrian H. Extremely slow Drude relaxation of correlated electrons. Nature, 438 (70/71);2005:1135–1137.

[25] Siegel R and Howell JR. Thermal Radiation Heat Transfer. Taylor & Francis, New York, NY; 2002.

[26] Godgaonkar DK, Varadan VV, and Varadan VK. Free space measurement of complex permittivity and complex permeability of magnetic materials at microwave frequencies. IEEE Transactions on Instrumentation and Measurements, 39 (2);1990:387–394.

[27] Kadaba PK. Simultaneous measurement of complex permittivity and permeability in the millimeter region by a frequency domain technique. IEEE Transactions on Instrumentation and Measurements, 33;1984:336–340.

[28] Baker-Jarvis J. Transmission/Reflection and Short-Circuit Line Permittivity Measurements. National Institute of Standards and Technology; July 1990.

[29] Weir W. Automatic measurement of complex dielectric constant and permeability at microwave frequencies. Proceedings of the IEEE, 62 (1);1974:33–36.

[30] Nicholson AM and Ross GF. Measurement of the intrinsic properties of materials by time-domain technique. IEEE Transactions on Instrumentation and Measurements, IM-19 (4);1970:377–382.

[31] Baker-Jarvis JR, Janezic MD, and Riddle BF, *et al.* Measuring the Permittivity and Permeability of Lossy Materials: Solids, Liquids, Metals, Building Materials, and Negative-Index Materials. NIST Technical Note 15362005; 2005.

[32] ASTM D2520-13, Standard Test Methods for Complex Permittivity (Dielectric Constant) of Solid Electrical Insulating Materials at Microwave Frequencies and Temperatures to 1650°C, ASTM International, West Conshohocken, PA; 2013. https://www.astm.org.

[33] Amirat Y and Shelukhin V. Homogenization of time-harmonic Maxwell equations and the frequency dispersion effect. Journal de Mathématiques Pures et Appliquées, 95;2011:420–443.

[34] MacKay TG. Homogenization of Linear and Nonlinear Complex Composite Materials. In: Weighofer WS and Lakhatakia A (eds.), Introduction to Complex Mediums for Optics and Electromagnetics; 2003. https://doi.org/10.1117/3.504610.ch14.

[35] Sihvola A. Electromagnetic mixing formulae and applications. IEE Electromagnetic Waves Series, 47;1999.

[36] Bossavit A. On the homogenization of Maxwell equations. International Journal for Computation and Mathematics in Electrical and Electronic Engineering, 14 (4);1995:23–26.

[37] El Feddi M, Ren Z, Razek A, and Bossavit A. Homogenization technique for Maxwell equations in periodic structures. IEEE Transactions on Magnetics, 33 (2);1997:1382–1385.

[38] Rayleigh L. On the electromagnetic theory of light. The London, Edinburgh, and Dublin Philosophical Magazine and Journal of Science, 12 (73);1967:81–101.

[39] Young A. Rayleigh scattering. Applied Optics, 20 (4);1981:533–535.

[40] Strutt JW. On the scattering of light by small particles. The London, Edinburgh, and Dublin Philosophical Magazine and Journal of Science, 41 (275);2009:447–454.

[41] Mie G. Beiträge zur Optik trüber Medien, speziell kolloidaler Metallösungen. Annalen der Physik, 330 (3);1908:377–445.

[42] Bohren CF and Huffmann DR. Absorption and Scattering of Light by Small Particles. Wiley-Interscience, New York, NY; 2010.

[43] Surjikov ST. Mie Scattering. A-to-Z Guide to Thermodynamics, Heat and Mass Transfer, and Fluids Engineering. Thermopedia.com, Retrieved 28 April, 2020.

[44] Balanis CA. Antenna Theory: Analysis and Design. John Wiley & Sons, Inc., Hoboken, NJ; 2005.

[45] Eugene FK. Radar Cross Section Measurements. Van Nostrand Reinhold, New York, NY; 1993.

[46] Shaeffer JF, Tuley MT, and Knot EF. Radar Cross Section. Artech House, Raleigh, NC; 1985.

[47] Meikle HD. Modern Radar Systems. Artech House, Morwood, MA; 2001.

Chapter 3
Antennas: properties, designs, and optimization

3.1 Introduction

The general discussion on propagation of electromagnetic waves in Chapter 2 outlines the physical background of nondestructive testing (NDT) using ground penetrating radar (GPR). The discussion was based on the electric and magnetic field intensities and the properties of materials, in which the electromagnetic waves propagate. Little has been said about how these waves are generated and coupled to the medium of interest; yet, the generation of electromagnetic waves and their reception are at the core of GPR. These waves are generated in radio-frequency circuitry, are amplified, then guided through transmission lines to the feed connection of an antenna that transmits the waves into the medium to be evaluated. Antennas are also responsible for the reception of waves scattered and reflected from artifacts in the medium (targets). The antenna transmits waves at proper levels and frequencies according to the GPR system specifications, quantities that are also related to the material properties of the target and host medium. In simple words, the antenna's goal is to transduce the power from an internal guided medium to the unbounded test medium.

As GPR's use proliferated and gained traction in diverse area of testing and evaluation, the demand for antennas specifically designed to address GPR requirements has grown rapidly due to the need for improved performance in different scenarios. The antenna is essential in guaranteeing the maximization of relevant parameters in GPR, such as resolution, depth of penetration, and target detection. Because of this, it is essential to understand how antennas operate in a given GPR scenario and the advantages of choosing one type of antenna over others. This can be done by a comprehensive investigation of the current distribution in the antenna and the antenna radiation pattern.

Different GPR applications require different types of antennas. For instance, hydrological applications require large depth of penetration with reduced resolution, whereas in inspection of concrete structures the depth is relatively low but resolution is critical for the detection of features within the concrete. Once the central frequency of the transmitted wave is defined, the choice of antenna also establishes the trade-off between resolution and depth of penetration. On the other hand, target identification usually imposes stricter requirements on antennas, compared to a GPR system aimed at merely target detection. In this case, the choice of the antenna can improve the quality of the assessment data, producing more accurate results. In many cases, weight and size of antennas play an important role in the planning of GPR surveys.

From an engineering point of view, the antenna can be considered as a linear system with its own transfer function, which is usually frequency dependent and with a nonlinear phase response. In general, most GPR antennas operate in a wide frequency range (wide bandwidth). For GPR applications, a common assumption is that antennas are made of perfect electric conductor materials, with possibly discrete lumped loads or radome enclosures distributed along and around their geometry. The antenna size is also related to the radiation pattern, with electrically small antennas having a broader radiation patterns than larger antennas.

In general, the antenna radiation characteristics are influenced by the host medium, properties which are specific to each survey. An example is antenna efficiency, which can be substantially reduced when it operates close to a dispersive medium. For antenna design purposes, realistic numerical models of GPR assessments are often required, enabling thus the understanding of how the antenna works, its interaction with the structure under test, and the establishment of its radiation properties.

The choice of the type of GPR system is also relevant to the considerations discussed here. Antennas for time-domain or impulsive GPRs are designed to avoid ringing (defined as parasitic oscillations that could hide signals reflected from targets). In these cases, nonlinear phase can be a critical consideration. For the case of frequency modulated or synthesized GPRs, the requirement for linear phase response from the antenna can be relaxed by performing initial calibration of the system.

The following sections discuss first the relevant parameters that characterize the radiation from antennas. Then, a set of frequently employed GPR antennas is briefly presented and explained in terms of these parameters. Finally, taking these geometries as a starting point of any antenna design, some recent GPR antenna designs are presented on the basis of computational optimization techniques.

It should also be noted here that the same antennas are used for reception of signals, either using a single antenna for both purposes or separate antennas for transmission and reception. It is not necessary to discuss receiving antenna separately because, based on the reciprocity principle, the parameters are the same, that is, the basic electric properties such as impedance, bandwidth, efficiency, and radiation pattern are the same for both functions. The obvious difference is in the power levels but since antennas are typically very efficient, power radiated has little effect on the structure of the antenna.

3.2 Antenna radiation parameters

Beyond the qualitative description provided in the introduction, the usefulness of antennas in GPR applications is derived from a set of standard quantitative parameters [1]. Taking into account the historical conception of these parameters, which was oriented toward telecommunication applications, formal definitions of antenna parameters [1] are being constantly updated to include the specific requirements to assess the quality of antennas in GPR surveys. For instance, there is still a lack of standardization in the definition of time-domain parameters for time-domain-radiated pulses [2–4]. To illustrate the definitions of the parameters and their meaning, this section is restricted to small antennas—often called Hertzian dipoles, electric dipoles, or elemental antennas. Deeper analysis can be made by considering the two basic canonic antennas: the thin-wire linear antenna (also called electric dipole) and the thin-wire circular loop (also called magnetic dipole), which also serve as references to all antennas used in practice since, for example, a long antenna or an antenna of any shape can be analyzed as an assembly of elemental antennas by superposition. Another good reason to consider elemental antennas is the availability of analytic or closed-form equations for the radiated fields, subject to their electrical size (e.g., the ratio of the minimum radius of a sphere containing the antenna over the wavelength of interest) [5,6]. Based on the availability of these closed-form expressions, electric and magnetic dipoles can be classified as infinitesimal dipoles, if the size of the dipole is very small (i.e., $\Delta l \ll \lambda$, where Δl is the physical length of the antenna and λ is the wavelength). For larger sizes but still smaller than the wavelength (i.e., $\Delta l < \lambda$), the antenna can be analyzed as an approximation to the small dipole. Antennas of the order of or larger than the wavelength (i.e., $\Delta l > \lambda$,) are analyzed as resonant antennas. The radiated fields from an infinitesimal linear antenna (electric or Hertzian dipole, Figure 3.1) can be calculated on the basis of these considerations. As long as an antenna geometry can be thought of as being composed of aggregated Hertzian dipoles, the fundamental radiation properties are illustrated properly by restricting the discussion to this particular kind of antennas.

To calculate the radiated fields of the Hertzian dipole, the dipole is assumed to carry a time-dependent current $I(t)$ and produces fields at a distance R from the antenna. The fields are due to the propagation of the wave over the distance R at a

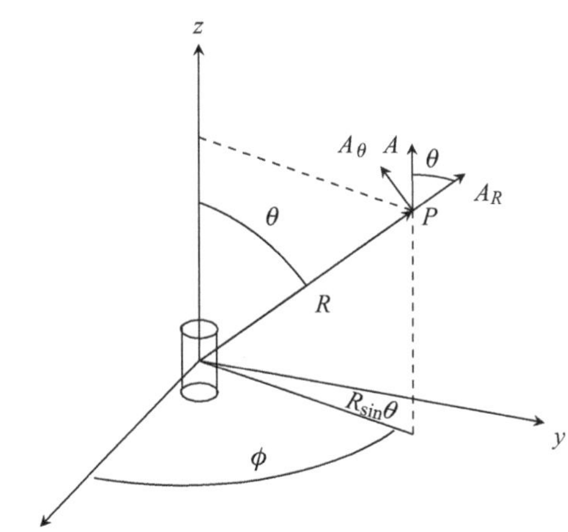

Figure 3.1 Hertzian dipole and the magnetic vector potential it produces shown in a spherical system of coordinates

phase velocity v_p. That is, a current at a time t produces a field at a distance R after a delay $\Delta t = R/v_p$. A more convenient way to view this delay due to the finite speed of propagation of the wave is to say that the field at a distance R, at a time t is due to a current at a time $t - \Delta t$. This is called a retarded current and may be written as

$$I = I_0 \cos\left(\omega t - \omega\frac{R}{v_p}\right) \quad [\text{A}] \tag{3.1}$$

where ω is the angular frequency of the time-dependent voltage source, R is the distance from the antenna, and I_0 is the amplitude of the current (peak value). ω/v_p is known as the phase constant β. To calculate the electric and magnetic field intensities in space, it is convenient to start with the magnetic vector potential. The magnetic vector potential for the Hertzian dipole of length Δl can be written as (see Figure 3.1)

$$\mathbf{A}(R,t) = \hat{z}\frac{\mu I_0 \Delta l}{4\pi R}\cos(\omega t - \beta R) \quad [\text{Wb/m}] \tag{3.2}$$

or, in phasor form, as

$$\mathbf{A}(R) = \hat{z}\frac{\mu I_0 \Delta l}{4\pi R}e^{-j\beta R} \quad [\text{Wb/m}] \tag{3.3}$$

From the vector potential in (3.3), it is a simple matter to obtain the magnetic field intensity from the definition of the vector potential: $\mathbf{B} = \mu\mathbf{H} = \nabla \times \mathbf{A}$. The use of the magnetic vector potential to define the magnetic field intensity is convenient since it is in the direction of the current (z-direction in Figure 3.1). Applying the curl on (3.3) and dividing by μ results in the magnetic field intensity in the wave:

$$\mathbf{H} = \hat{\phi}\frac{I_0 \Delta l \beta^2}{4\pi}e^{-j\beta R}\sin\theta\left(\frac{1}{j\beta R} + \frac{1}{(j\beta R)^2}\right) \quad [\text{A/m}] \tag{3.4}$$

The electric field intensity is obtained from Ampere's law (see (2.18)), which in space, where there are no sources of current, is

$$\nabla \times \mathbf{H} = j\omega\varepsilon\mathbf{E} \tag{3.5}$$

Performing the curl operation on (3.4), we obtain the electric field intensity at a distance R from the antenna:

$$\begin{aligned}\mathbf{E} = &-\hat{R}\frac{\eta I_0 \Delta l \beta^2}{2\pi}e^{-j\beta R}\cos\theta\left(\frac{1}{(j\beta R)^2} + \frac{1}{(j\beta R)^3}\right)\\ &-\hat{\theta}\frac{\eta I_0 \Delta l \beta^2}{4\pi}e^{-j\beta R}\sin\theta\left(\frac{1}{j\beta R} + \frac{1}{(j\beta R)^2} + \frac{1}{(j\beta R)^3}\right) \quad [\text{V/m}]\end{aligned} \tag{3.6}$$

where $\eta = \sqrt{\mu/\varepsilon}$ is the intrinsic or wave impedance of the surrounding medium and is also called the wave impedance. Since the electric and magnetic field intensities are composed of terms that decrease as $1/R$, $1/R^2$, and $1/R^3$, three distinct domains can be defined on the basis of which of the three terms dominates [5]. The first is the zone in which the term $1/R^3$ dominates, corresponding to small values of R, named the near-field reactive zone followed by the zone in which the term $1/R$ dominates, for large values of R, called the far-field or Fraunhofer zone. An intermediate zone, in which the term $1/R^2$ dominates, is also defined formally but it is usually viewed as a transition between the near and far fields and is of little importance in the behavior of antennas. It is called an induction zone because of the form of the equations that contain the term $1/R^2$. The transition zone is also called the radiating near field or Fresnel zone to distinguish it from the near field in which radiation cannot occur, as will be seen shortly. Although many antenna parameters, including those discussed here, are defined on the basis of far-field behavior, GPR is often used in the radiating near field where the parameters apply as well.

For Hertzian dipoles, the near-field region is any location $R \ll \lambda$, with λ the wavelength of propagation. In this zone, the terms in the electric field intensity containing $1/R$ and $1/R^2$ are negligible compared to the term containing $1/R^3$ and are therefore neglected. Because $R \ll \lambda$, and since $\lambda = 2\pi/\beta$, the term $e^{-j\beta R}$ equals approximately 1 and, therefore, the electric field intensity behaves as that of an electrostatic dipole. Using the same arguments, in the expression for the magnetic field intensity the term containing $1/R$ is small compared to that containing $1/R^2$ and is therefore neglected. The electric and magnetic field intensities in the near field become

$$\mathbf{H} = \hat{\phi}\frac{I_0\Delta l}{4\pi R^2}\sin\theta \quad [\mathrm{A/m}] \tag{3.7}$$

$$\mathbf{E} = -\hat{R}j\frac{\eta I_0\Delta l}{2\pi\beta R^3}\cos\theta - \hat{\theta}j\frac{\eta I_0\Delta l}{4\pi\beta R^3}\sin\theta \quad [\mathrm{V/m}] \tag{3.8}$$

Since the electric field intensity in the near field is purely imaginary, whereas the magnetic field intensity is real, the time-averaged power density in the near field is zero (that is, the Poynting vector is purely imaginary), and no propagation of electromagnetic waves can occur. In short, the dipole does not radiate power and the near field is characterized by storage of energy. Because the electric field intensity is much larger than the magnetic field intensity, the stored electric energy density is higher than the stored magnetic energy density, meaning that the Hertzian dipole in the near field is essentially capacitive in nature.

In the far-field zone, where the distance R is large compared to the wavelength ($R \gg \lambda$), the terms containing $1/R^2$ and $1/R^3$ in (3.4) and (3.6) can be neglected, and the magnetic and electric field intensities reduce to the following:

$$\mathbf{H} = \hat{\phi}\frac{jI_0\Delta l\beta}{4\pi R}e^{-j\beta R}\sin\theta \quad [\mathrm{A/m}] \tag{3.9}$$

$$\mathbf{E} = \hat{\theta}\frac{j\eta I_0\Delta l\beta}{4\pi R}e^{-j\beta R}\sin\theta \quad [\mathrm{V/m}] \tag{3.10}$$

The following characteristics of the far-field zone can be inferred by inspection of these equations:

- The electric and magnetic field intensities are perpendicular to each other.
- The direction of propagation is given by the Poynting vector and is orthogonal to the **E** and **H** fields.
- The fields in the far-field domain decrease as $1/R$.

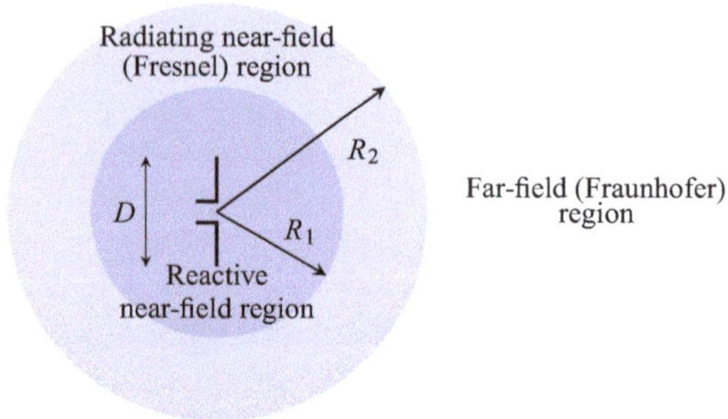

Figure 3.2 Field zones of an antenna

- The ratio between the amplitude of the electric and magnetic field intensities is equal to the intrinsic impedance of the medium η, hence this ratio is called radiation impedance.

The fields in the far field are fundamentally different than in the near field not only in that they propagate power but also because they interact with media differently. This includes attenuation, reflection, refraction and, scattering of waves. Although the focus in this chapter is the antennas themselves, it is important to keep in mind that the purpose of antennas is to couple power into the surrounding media and the fields associated with that power are those characteristic of the zone in which the media are located.

Practical antennas, including GPR antennas, are almost always larger in size than Hertzian dipoles, and targets in surveys can be located in the near-, intermediate-, or far-field zones, as illustrated in Figure 3.2. Note the distinction between the reactive near field in which there can be no radiation and the radiating near field in which some radiation may occur but it is small compared to the far field. To handle the very different behaviors in the three regions, the boundaries between zones can be handled using simple relations [5], but the properties presented for Hertzian dipoles are valid. The exact expressions for arbitrary antennas are sometimes difficult to obtain and the properties of these antennas will vary but only in values, not in the fundamentals. For example, the electric or magnetic fields may have different amplitudes but they still decrease as $1/R$ in the far field. For this reason, for definition and basic analysis purposes, antenna parameters of the Hertzian dipole in the near- and far-field regions are sufficient for the purpose of this section. These properties are described next and, aside from the exact expressions shown for the Hertzian dipole, these are the properties one has to consider for any antenna.

3.2.1 Radiated power

The time-averaged power density radiated by the antenna in the far field is defined by the Poynting vector from the electric and magnetic field intensities. Using (3.9) and (3.10) for the Hertzian dipole:

$$\mathcal{P}_{av} = \frac{1}{2}Re\{\mathbf{E} \times \mathbf{H}^*\} = \hat{R}\frac{\eta I_0^2 (\Delta l)^2 \beta^2}{32\pi^2 R^2} \sin^2\theta \quad [\mathrm{W/m^2}] \tag{3.11}$$

The total radiated power can be calculated by surrounding the antenna with a sphere of radius R and integrating the radiated power density over the sphere to derive the total power traversing the surface of the sphere. From (3.11) in spherical coordinates

$$\mathbf{P}_{rad} = \int_s \mathscr{P}_{av} \cdot ds = \int_0^{\phi=2\pi} \int_0^{\theta=\pi} \frac{\eta I_0^2 (\Delta l)^2 \beta^2}{32\pi^2 R^2} \sin^2\theta R^2 \sin\theta d\theta d\phi \quad [\mathrm{W}] \tag{3.12}$$

Integrating gives

$$\mathbf{P}_{rad} = \frac{\eta I_0^2 (\Delta l)^2 \beta^2}{12\pi} = I_0^2 \frac{\eta\pi (\Delta l)^2}{3\lambda^2} \quad [\mathrm{W}] \tag{3.13}$$

where $\beta = 2\pi/\lambda$ was used to obtain the second form on the right-hand side of (3.13) leading to a simplified form as well as showing explicitly that power radiated depends on the square of the ratio between antenna length and wavelength. Radiated power of the infinitesimal dipole is also proportional to the current squared and the intrinsic impedance of the medium in which the antenna radiates. It should be noted that the radiated power calculated here is independent of the distance from the antenna as a consequence of the fact that any losses in the medium have been neglected. Note as well that the radiated power density (Equation (3.11)) is dependent on distance indicating the spread of power over larger and larger areas as the distance increases.

3.2.2 Antenna radiation patterns

In general, the electric and magnetic fields in the far-field region are dependent on the angles θ, ϕ for any sphere of radius R, as can be seen in Figure 3.1. That is, in the far-field region where fields depend on $1/R$, the electric and magnetic fields radiated by the antenna are direction dependent. Equation (3.11) also shows that the time-averaged power density radiated by the antenna is position dependent, specifically, in this case depending only on θ. When the expressions for the electric field intensity, magnetic field intensity, and power density are complex, it is useful to be able to describe their dependence in space in a simple fashion. This is done through the radiation patterns of the antenna. The radiation pattern is a plot of the electric or magnetic field intensity or the power density as a function of the space coordinates, either as a three-dimensional (3D) plot in space or as a two-dimensional (2D) plot on any convenient plane. In simple terms, radiation patterns show the distribution of fields or power density in space. Radiation patterns provide a simple, graphical means of displaying the radiation characteristics of the antenna.

There are two types of radiation patterns; field radiation patterns describing the distribution of the electric or magnetic field intensity, and power radiation patterns, describing the distribution of the radiation power density. These are discussed next in some detail.

3.2.2.1 Antenna field radiation patterns

An antenna field radiation pattern is defined as the relative (or normalized) magnitude of the electric or magnetic field intensity at any distance from the antenna in the far field. In general, radiation patterns are presented in spherical coordinates, and they require a 3D plot. 3D plots are difficult to execute and very often difficult to interpret. For representation purposes, it is common to represent polar or rectangular plots as planar projections (such as on $\phi = 0$ or $\theta = \pi/2$) from the 3D pattern.

For instance, in Hertzian dipoles, the electric field intensity in (3.10) depends only on θ and R. For orientation purposes, the vertical plane containing the dipole is called the E-plane, whereas the horizontal plane perpendicular to the dipole is called the H-plane. Since one can plot the pattern of the electric or magnetic field intensity either in the E-plane or H-plane, there are four types of planar radiation patterns that are commonly referred to. These are the E-field E-plane and H-field E-plane radiation patterns showing the pattern of the electric and magnetic field intensities in the plane containing the dipole ($\phi = 0$), and similarly, the E-field H-plane and H-field H-plane radiation patterns showing the pattern of the electric and magnetic field intensities in the plane perpendicular to the dipole ($\theta = \pi/2$). One can of course define a radiation pattern by cutting an arbitrary plane through the 3D pattern simply by substituting specific values for ϕ and θ, and, for more complex antennas, that may be useful. For simple, straight wire antennas the previous four patterns are sufficient. In most cases, the electric field pattern is sufficient to describe the field distribution of the antenna and for this reason, unless specifically indicated, when a field radiation pattern is given, it is understood to be an E-field pattern. For either type of field pattern, the planes on which they are obtained must be specified.

To define the electric field pattern, we start with the magnitude of the electric field intensity of the dipole:

$$|E| = \left|\frac{j\eta I_0\beta\Delta l}{4\pi R}e^{-j\beta r}\sin\theta\right| = \left|\frac{j\eta I_0\beta\Delta l}{4\pi R}\right||\sin\theta| \tag{3.14}$$

Normalized patterns of (3.14) in terms of the amplitude of E are usually depicted by dividing the equation by the magnitude of the field. For instance, the normalized E-field, E-plane ($\phi = 0$) radiation pattern of the Hertzian dipole is

$$|f_e(\theta)| = |\sin\theta| \tag{3.15}$$

Similarly, the E-field, H-plane ($\theta = \pi/2$) radiation pattern is

$$|f_e(\phi)| = 1 \tag{3.16}$$

The two patterns are shown in Figure 3.3 in polar representation over the respective projection planes.

3.2.2.2 Antenna power radiation pattern

The considerations of the previous section apply equally to the radiated power density of the antenna in (3.11). We first write the magnitude of the power density as

$$|\mathbf{P}_{av}| = \left|\hat{R}\frac{\eta I_0^2(\Delta l)^2\beta^2}{32\pi^2R^2}\right||\sin^2\theta|\,[\mathrm{W/m^2}] \tag{3.17}$$

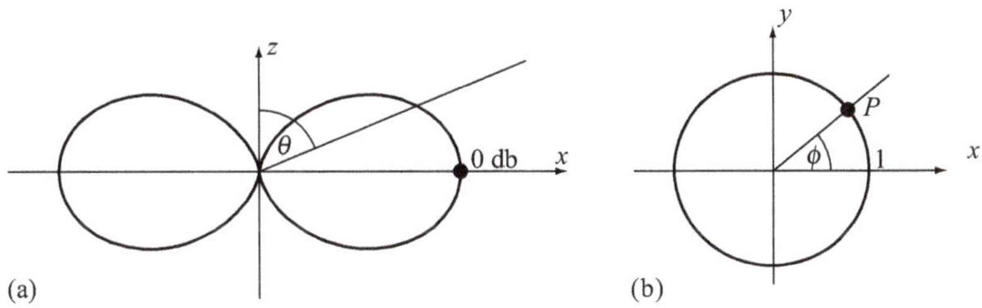

Figure 3.3 (a) Normalized electric field radiation pattern in the E-plane and (b) normalized electric field radiation pattern in the H-plane

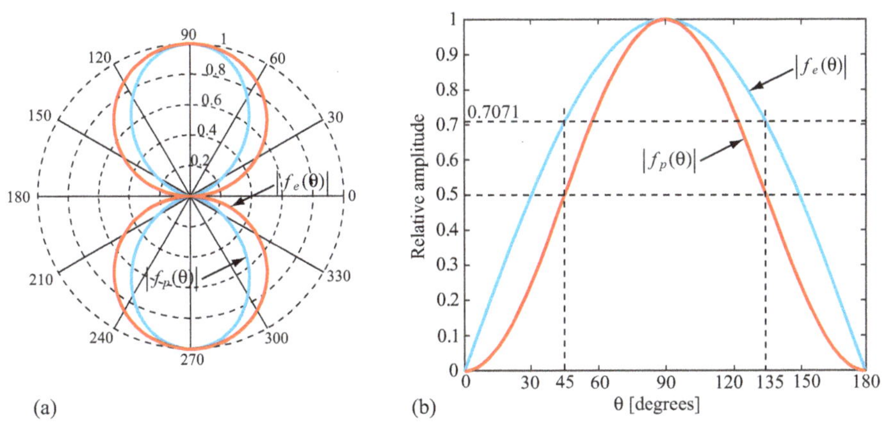

Figure 3.4 (a) Normalized field and power radiation patterns in the E-plane in polar coordinates and (b) normalized field and power radiation patterns in the E-plane in rectangular coordinates

The power radiation pattern is obtained by dividing the magnitude of the amplitude

$$|f_p(\theta)| = |\sin^2\theta| \tag{3.18}$$

Note that the power radiation pattern is the square of the field radiation pattern $[|f_p(\theta)| = (|f_e(\theta)|)^2]$. This relation is not unique to Hertzian dipoles and holds for any antenna. Figure 3.4 shows the normalized field radiation pattern and power radiation pattern for the Hertzian dipole in polar and rectangular representations. The latter is sometimes more useful and easier to interpret than the polar form.

3.2.2.3 Beamwidth

For any radiation pattern, the beamwidth (BW) is defined as the angle between the two points of the radiation pattern where the beam is 3 dB below its maximum value (half-power directions) or, equivalently, where the field is $1/\sqrt{2}$ of its peak value.

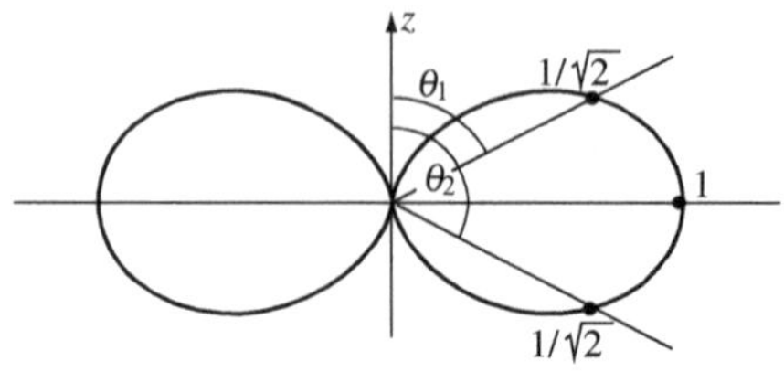

Figure 3.5 Beamwidth shown on the electric field intensity normalized radiation pattern on the E-plane in polar representation

Figures 3.4 and 3.5 show the concept. In Figure 3.5, the BW is $\Delta\theta = \theta_2 - \theta_1$. When the pattern is more complex, it may be more difficult to ascertain the BW but the concept is clear. Note also that in the H-plane, the field distribution is uniform, meaning that the radiation in this plane is omnidirectional. Again, by way of example, the BW of the Hertzian dipole is 90° (between the half-power points at 45° and 135°) as can be seen in the radiation pattern plots in Figure 3.4(b).

3.2.3 Radiation intensity

Radiation intensity is defined as the time-averaged power of the antenna per unit solid angle, given in watts per steradian [W/sr]. Since the total radiated power is integrated over 4π unit solid angles, the radiation intensity may be written as

$$U(\theta,\phi) = \mathscr{P}_{av}R^2 \quad [\mathrm{W/sr}] \tag{3.19}$$

where the dependency on θ and ϕ is added to indicate that, in general, the radiation intensity depends on both angles as it does for the time-averaged Poynting vector, $\mathscr{P}_{av}$. In the case of the Hertzian dipole, the radiation intensity is

$$U(\theta) = \frac{\eta I_0^2\beta^2(\Delta l)^2}{32\pi^2}\sin^2\theta = \frac{\eta I_0^2(\Delta l)^2}{8\lambda^2}\sin^2\theta \quad [\mathrm{W/sr}] \tag{3.20}$$

Radiation intensity is independent of the distance from the dipole and only depends on θ. Clearly the radiation intensity is not uniform over a sphere of any radius. An average radiation intensity is defined by spreading the total power radiated uniformly over the area of the sphere of radius R. Since there are 4π steradians, this is simply equal to $P_{rad}/4\pi$. More formally, one can write

$$U_{av} = \frac{\int_s U(\theta,\phi)\,ds}{4\pi} = \frac{\int_s P_{av}R^2 d\Omega}{4\pi} = \frac{P_{rad}}{4\pi} \quad [\mathrm{W/sr}] \tag{3.21}$$

The average radiation intensity is viewed as the radiation intensity of an isotropic source, with a total radiated power equal to that of the antenna under analysis. The concept of an isotropic source, which does not exist in reality, is a convenient quantity to use in computation of antenna parameters and, in particular, when trying to compare antennas with different radiation patterns. For the Hertzian dipole, the average radiation intensity is

$$U_{av} = I_0^2 \frac{\eta(\Delta l)^2}{12\lambda^2} \quad [\mathrm{W/sr}] \tag{3.22}$$

The radiation intensity and average radiation intensity are parameters in their own right but they are also useful in defining properties of even more consequence, including antenna directivity and antenna gain that are discussed in the following two sections.

3.2.4 Antenna directivity

Except for the isotropic radiator, the power density of any antenna varies in space from point to point. For this reason, it is said that an antenna has directive properties, which are characterized through a dimensionless parameter called the directivity of the antenna. The directivity is defined as the ratio between the radiation intensity in a given direction (θ, ϕ) and the averaged radiation intensity:

$$D(\theta, \phi) = \frac{U(\theta, \phi)}{U_{av}} = \frac{4\pi U(\theta, \phi)}{P_{rad}} \tag{3.23}$$

Note that if the antenna radiates uniformly in all directions in space, the average radiation intensity and the radiation intensity are equal and directivity is 1 everywhere. This is a truly omnidirectional radiator but this can only occur for isotropic radiators, which, as indicated earlier, cannot be built in practice. For all other cases, the directivity varies from point to point. The maximum directivity is more often the parameter of interest. Directive antennas are the type of interest in GPR systems, although the exact directivity depends on the type of antenna and the application. The directivity of the Hertzian dipole is obtained by dividing (3.20) by (3.22):

$$D(\theta) = \frac{3}{2} \sin^2 \theta \tag{3.24}$$

The dipole radiates maximum power density in directions where ($\sin^2 \theta = 1$), whereas in the direction of the dipole (z axis), the radiated power density is zero. Maximum directivity of the Hertzian dipole is 1.5, as can be seen from (3.24). Note also that a feel for (but not values of) directivity can be obtained from the radiation pattern (see e.g., Figure 3.4 or 3.5). Perhaps the simplest way to think about directivity is that it indicates the power density radiated in any direction in space compared to the power density radiated by an isotropic radiator that radiates the same total power.

3.2.5 Antenna gain

The directivity as given in (3.23) or (3.24) disregards any losses that may have occurred in the antenna. Instead of relating the averaged radiation intensity in terms of the radiated power, the input power to the antenna can be considered, including the losses in the antenna. The antenna gain (also called power gain) is defined as the ratio between the maximum radiation intensity and the average radiation intensity of a perfect isotropic radiator with the same input power:

$$G(\theta, \phi) = \frac{U(\theta, \phi)_{\max}}{P_{in}/4\pi} = \frac{4\pi U(\theta, \phi)_{\max}}{P_{in}} \quad (3.25)$$

From its definition, it is obvious that antenna gain has the same meaning and use as maximum antenna directivity, but it is more accurate in the sense that it takes into account the power loss in the antenna itself. For high-efficiency antennas it is not uncommon to use maximum directivity and gain interchangeably.

3.2.6 Polarization

Polarization is the pattern in space described by the instantaneous electric field intensity vector. In the general case, this vector describes an elliptical helix in the direction of propagation. Other figures or patterns may be possible such as linear and circular polarizations. Polarization is a relevant parameter in a transmit/receive system, such as GPR, because the performance of the system depends on the polarization of antennas. As examples, circular polarized antennas can receive any linear or elliptical polarized waves, whereas linearly polarized antennas can only fully receive linearly polarized waves, and even waves from linearly polarized antenna may not be received from other perpendicularly linear polarized antennas. Power loss due to these effects is due to polarization mismatch between antennas. It should be remembered that the colloquial use of antenna polarization refers to the polarization of the wave (transmitted or received), and not to any concept of physical orientation of antennas. Thus, for example, a linearly polarized antenna means that the antenna transmits or can receive linearly polarized waves.

Commercial GPR antennas can be found with linear and circular polarization, depending on the application, with the aim to minimize polarization mismatch. However, polarization mismatch is not totally avoidable in GPR because the radiated field is distorted when propagating in the medium under test. In general, GPR surveys depend on both the target and the antenna orientation. For vertical targets and linear antennas with transverse orientation, the received signal can be very weak. To overcome this drawback, cross-polarized antennas (e.g., two linear antennas perpendicular to each other) are employed in order to detect simultaneously reflections from target in two orthogonal directions.

3.2.7 Radiation resistance

In analogy with the dissipated power in a resistor, one can define an equivalent resistance to account for the radiated power as

$$P_{rad} = I_{rms}^2 R_{rad} = \frac{I_0^2}{2} R_{rad} \qquad [\mathrm{W}] \quad (3.26)$$

where R_{rad} is called the radiation resistance of the antenna, and $I_{rms} = I_0/\sqrt{2}$ is the root mean square of the feeding current. Radiation resistance simply indicates that the power the dipole can radiate for a given current. Maximization of the radiation resistance means that the antenna can radiate more power for a given current.

For the Hertzian dipole, the radiation resistance is (using (3.13) for radiated power)

$$R_{rad} = \frac{2\eta\pi}{3}\left(\frac{\Delta l}{\lambda}\right)^2 \quad [\Omega] \tag{3.27}$$

Equation (3.27) shows that the power radiated by an antenna of this type is directly proportional to the square of the length of the antenna, Δl, and inversely proportional to the square of the wavelength. However, this is a consequence of the assumption that $\Delta l \ll \lambda$ and is specific to very short dipoles. It does not hold in general. Antennas also have reactance and in fact the reactance can be significant. In that sense, one should properly talk about radiation impedance. However, since the important parameter is the radiated power, and that depends on the radiation resistance of the antenna, it is uncommon to specify radiation impedance. (see, however, the discussion in the following section).

3.2.8 Input impedance

An antenna can be thought of as a circuit element connected by means of a transmission line to a generator, which supplies the power radiated by the antenna into unbounded space in the general case or into a medium in the case of GPR (Figure 3.6). The equivalent circuit in Figure 3.7 can then be used to characterize the power transmitted. Viewing the antenna as a load, its input impedance is $Z_{in} = R_{rad} + R_{ant} + jX_{ant}$ where R_{rad} is the radiation resistance defined in (3.27), R_{ant} is the physical (ohmic) resistance

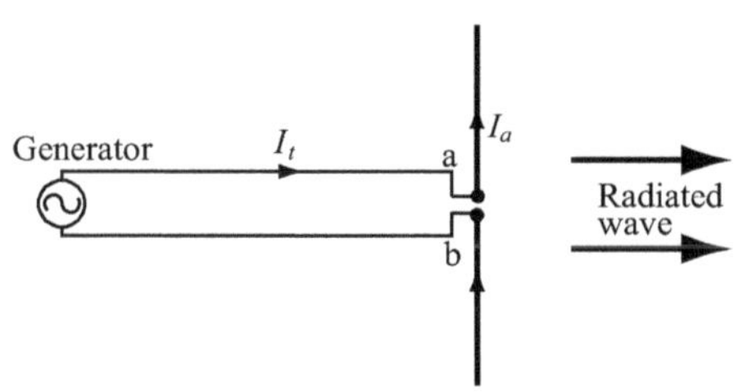

Figure 3.6 A voltage generator connected to a dipole antenna by means of a transmission line

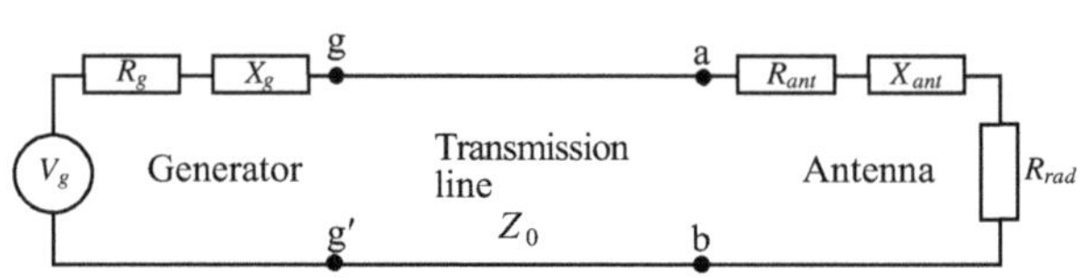

Figure 3.7 Equivalent circuit representation of the generator, transmission line, and antenna

of the element that leads to power losses in the antenna, and X_{ant} is the reactance of the antenna. Assuming the transmission line is matched at the generator and load, the generator impedance appears at the antenna connections, and the power radiated is

$$P_{rad} = \frac{I_a^2 R_{rad}}{2} = \frac{1}{2} \frac{V_g^2 R_{rad}}{(R_{ant} + R_{rad} + R_g)^2} \quad [\mathrm{W}] \tag{3.28}$$

Proper operation of any antenna usually requires that it be matched, either by guaranteeing minimum reflection (low voltage standing wave ratio (VSWR) on the line) or conjugate matching (whereby $jX_g = -jX_{ant}$ and $R_g = R_{rad} + R_{ant}$). Under these conditions, the radiated power is

$$P_{rad} = \frac{1}{2} \frac{V_g^2 R_{rad}}{(2R_{rad} + R_{ant})^2} \quad [\mathrm{W}] \tag{3.29}$$

In most practical antennas the ohmic resistance is small compared to the radiation resistance and, as a first approximation, may be neglected. Doing so gives the maximum radiated power:

$$P_{rad\text{-}\max} = \frac{V_g^2}{8R_{rad}} \quad [\mathrm{W}] \tag{3.30}$$

Inspection of (3.29) indicates that the losses in the antenna are due to its ohmic resistance. This allows us to estimate the losses in the antenna and hence the radiation efficiency of the antenna as

$$eff = \frac{R_{rad}}{R_{rad} + R_{ant}} \times 100\% \tag{3.31}$$

There are of course other sources of losses and therefore the efficiency in (3.31) should be viewed as a high estimate.

3.2.9 Bandwidth

Bandwidth accounts for the range of frequencies over which the antenna maintains radiation parameters within minimal, acceptable deviations. In GPR systems, the bandwidth is a paramount parameter because antennas with narrow bandwidths (such as resonant antennas) cannot be used to transmit pulses with large frequency content (narrow pulses). On the other hand, a wide bandwidth is often desirable as it improves performance. In terms of the input impedance, the real-valued VSWR is used to account for changes in amplitudes as a function of frequency. Relative bandwidth is calculated using the relation:

$$B = \frac{\Delta f}{f_c} \times 100\% \tag{3.32}$$

where $\Delta f = f_u - f_l$, f_u is the upper frequency in the band and f_l is the lower frequency. The central frequency is $f_c = (f_u + f_l)/2$. For broadband applications, the so-called $x:y$ notation is used. For instance, a 3:1 bandwidth means that the upper frequency is three times higher than the lower.

Antennas capable of transmitting (and receiving) over a wide bandwidth are called broadband antennas and are distinctly different than resonant antennas. The latter, including the Hertzian dipole, are narrowband antennas and behave as narrow band-pass filters. Some broadband antennas will be discussed separately in Section 3.4 in the context of antennas for GPR.

3.2.10 Pulse fidelity

Closely related to the bandwidth, pulse fidelity is the time-domain parameter that relates the current and voltage at the feed point [7]. Mathematically, pulse fidelity F is the maximum cross-correlation between the normalized output pulse, namely, the current $i(t)$, and the reference input voltage pulse $v(t)$. The value of F is equal to one (maximum value), when the input and output pulses are identical in shape. Cross-correlation delay time may be used during any optimization of F, thus removing the effect of nonzero time delay between waveforms. Pulse fidelity and its optimization are commonly used in impulse GPR systems, where both the pulse shape and pulse length have an effect on resolution of the system.

3.2.11 Group delay

Any signal transmitted by a radiating antenna experiences a time delay called group delay. It is a measure of time distortion in the pulse due to the frequency dependence of the phase of the pulse as it propagates in media. The group delay is calculated as

$$\tau_g = -\frac{d\Phi(\omega)}{d\omega} \text{ [s]} \tag{3.33}$$

where $\Phi(\omega)$ is the phase, which in general is frequency dependent, ω is the angular frequency, and τ_g is the group delay. A constant group delay versus frequency leads to a constant propagation time delay and an undistorted output signal. The larger the group delay the higher the distortion in the pulse with consequences on performance of the system. This quantity is directly related to the concept of dispersion, whereby different frequency components of a pulse arrive at the target (or back at the receiving antenna) at different times and hence distort the pulse (see discussion in Section 2.4).

In GPR, group delay is related to the size and depth of the target. The time delay between transmitted and received pulses determines the distance between source and target. A linear relation between phase and frequency leads to a constant phase velocity and hence a linear group delay. To properly account for depth of targets, it is imperative to be able to correct the velocity of the wave using the group delay to account for the different velocity of propagation in host media. A frequency-dependent group delay influences the assessment of depth and also limits the resolution because of the widening of the radar pulse in time. For multiple, closely spaced objects, an impulse train of pulses is produced at the receiving antenna. Non-dispersed pulses (i.e., pulses that do not experience group delays) are easier to separate as their amplitude decreases to zero before the subsequent pulses arrive. By contrast, dispersed pulses are wider in time, and dispersion causes overlap between adjacent pulses. Whereas group delays caused by propagation in media cannot be avoided, GPR antennas should be designed to minimize group delay variations over the radiating bandwidth due to the antenna and its circuitry.

3.2.12 Receiving antenna parameters

Properties of antennas in receiving and transmitting modes are entirely equivalent, that is, the antenna parameters discussed in the previous sections apply to receiving antennas as well. This remarkable aspect of antennas is assured by the reciprocity theorem. Figure 3.8 shows two antennas (not necessarily of the same type, dimensions or orientation) in transmitting/receiving mode. The reciprocity theorem states that if a source is applied to antenna 1 and a signal is received in antenna 2, then, applying the same identical source to antenna 2 will produce an identical signal in antenna 1. This theorem only applies if the properties of the material between the two antennas are the same in either direction (isotropic materials), which is normally the case in free space or, as it happens in most host media encountered in GPR, and in most dielectrics. The reciprocity theorem is a fundamental theorem in communication systems, of which GPR is clearly a part, even though it does assume isotropy. In GPR assessments, one may from time to time encounter anisotropy and, in spite of the fact that it may be mild, anisotropy affects the performance of GPR as will be discussed in the following chapter. To illustrate the theorem, consider a transmitting Hertzian dipole antenna producing a plane wave in the far-field region (see (3.9) and (3.10)). At some position in the far field, another Hertzian dipole acts as receiving antenna (Figure 3.9). The

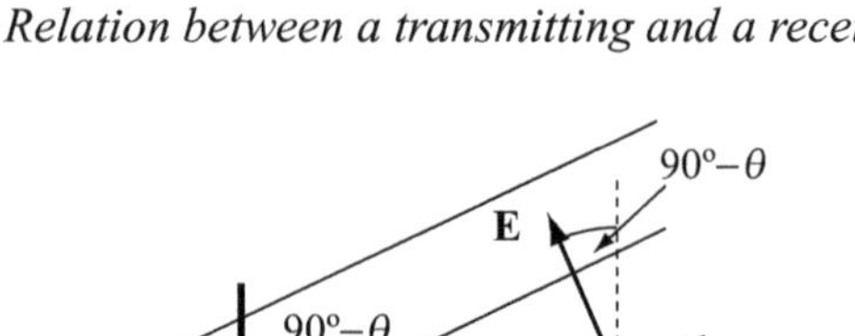

Figure 3.8 Relation between a transmitting and a receiving antenna

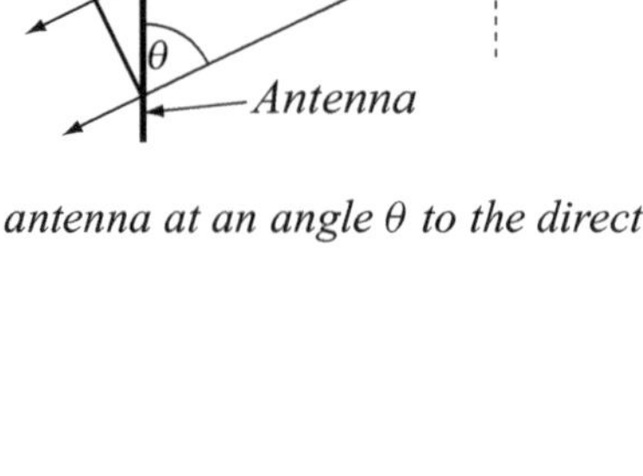

Figure 3.9 A receiving antenna at an angle θ to the direction of propagation of a plane wave

receiving antenna is shown at an angle θ to the direction of propagation of the wave. The open circuit voltage in the receiving dipole is

$$V_a = \mathbf{E} \cdot \Delta \mathbf{l} = E\Delta l \cos(90 - \theta) = E\Delta l \sin\theta \qquad [\mathrm{V}] \tag{3.34}$$

Regarding the receiving properties, the most relevant fact in (3.34) is that the received voltage is proportional to $\sin\theta$. Therefore, from the representation of radiation patterns, one infers that the receiver and transmitter radiation patterns are exactly the same, and that is predicted by the reciprocity theorem. That means as well that other properties such as directivity and gain are the same. Properties such as radiated power and power density depend on the current in the antenna and are irrelevant for receiving antennas. However, the reciprocity theorem indicates that the radiated power and power density obey the same expressions. Similarly, identical antennas will have the same radiation resistance in either mode.

3.2.13 Effective aperture

Only a small percentage of the power transmitted by a transmitting antenna impinges on (is intercepted by) a receiving antenna. Since the parameter of importance at the receiving antenna is the power density (see (3.11)), the power received by the antenna is proportional to the effective area or effective aperture of the receiving antenna. The use of the qualification "effective" indicates that since the antenna receives power, it must have an effective area even if the physical antenna does not have an area, as in the case of the Hertzian dipole and many other antennas. To understand the concept, consider any source antenna (an isotropic or uniformly radiating antenna is most convenient for this purpose because of its simplicity) that radiates a total power P_{rad} as in Figure 3.10. At a distance R from the source, the power density is $\mathscr{P}_{av} = P_{rad}/4\pi R^2$. The effective area of an antenna (A_a) is the ratio between the total time-averaged power received by the antenna P_{ant} and the time-averaged power density at the location of the antenna. Effective area is therefore a parameter that specifies the ability to collect power in a receiving antenna regardless of its actual geometry. Although the isotropic antenna was used to define the concept, the result applies to any antenna.

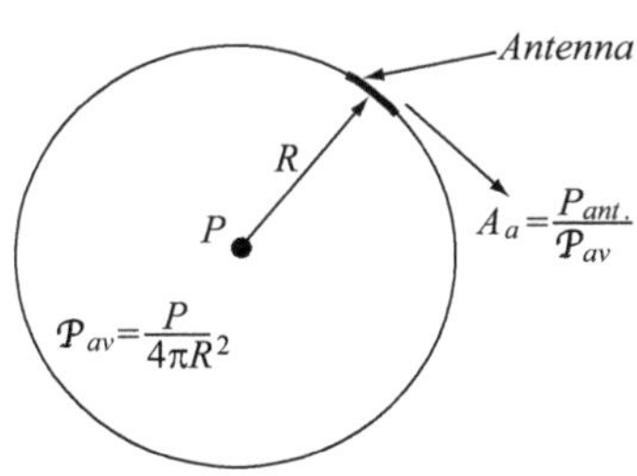

Figure 3.10 Relation between effective area of an antenna and power received

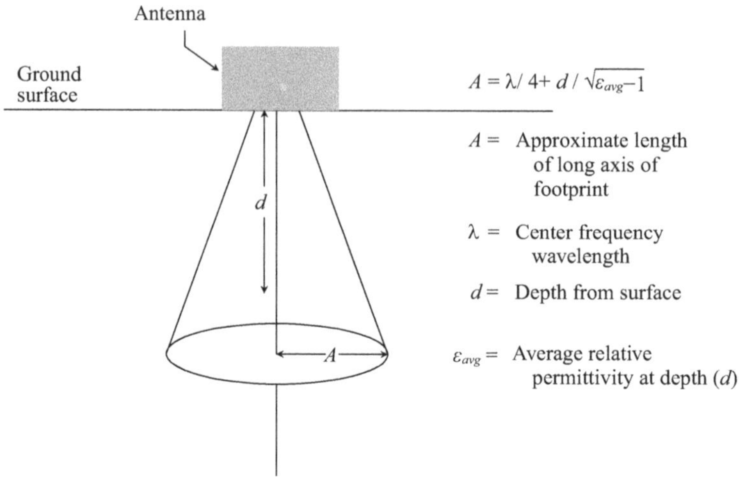

Figure 3.11 Antenna footprint

3.2.14 Antenna footprint

The antenna footprint, illustrated in Figure 3.11, is defined as the illuminated area in the horizontal plane at a given distance from the antenna in the structure under test. This energy projection is highly dependent on frequency, material properties, and antenna height [8]. The antenna footprint is therefore a trade-off between these three parameters as well as antenna size and geometry (see also the discussion in Section 1.3.3). If the transmitted power of the antenna is spread across an area larger than the target size, the backscattered pulse will be significantly weaker than the reflected pulse from the surface. For closely spaced targets, a small footprint is highly desirable as it contributes to an increase in the resolution of the system. In addition, small footprint is useful to avoid unnecessarily large surface reflections and clutter. The optimal size and shape of the footprint can vary depending on the GPR application. For instance, the detection of large metallic bars in concrete structures requires larger footprints than the detection of small targets in the same host medium, such as cracks or inclusions. Similar considerations applied to the detection of pipelines are archaeological features.

3.3 Antenna interaction with the medium under test

In general, antenna characteristics are also influenced by the host medium not the least due to the proximity of antennas to the host medium's surface and the medium's lossy nature. For example, when the antenna height is small compared to the skin depth in the lossy material, the input resistance is higher than its free-space values [5]. Radiation patterns and footprints are also affected by the constitutive parameters of the host media at low-height antenna cases. Physically, time-varying EM fields radiated by the antenna generate time-dependent currents along the host surface and within a few skin depths that, in turn, act as sources of scattered electromagnetic fields. These interfere with the primary antenna currents and, as a result, both the electromagnetic field distribution in space and the input impedance, and consequently the resonant frequency, are altered in comparison with the free-space case.

A simplified procedure of analysis of interaction of antennas with media, based on the plane wave approximation and geometric optic models [9], is illustrated in Figure 3.12. In general, the simplest scenario for GPR is an antenna over a dielectric half space. Radiated fields are refracted in the dielectric according to the boundary conditions resulting from the slower velocity of propagation in the dielectric. As a result, the directivity and radiated waveform are modified.

Using Figure 3.12, the total air wave E^{air} from an infinitesimal dipole at a height h cam be calculated as the sum of the direct and reflected air waves $E^{air} = E^{air}_{direct} + E^{air}_{reflect}$. The direct air wave is

$$E^{air}_{direct} = E_0 \frac{h}{h + \sqrt{\varepsilon_r} d} e^{-jk_0(\sqrt{\varepsilon_r} d)} \tag{3.35}$$

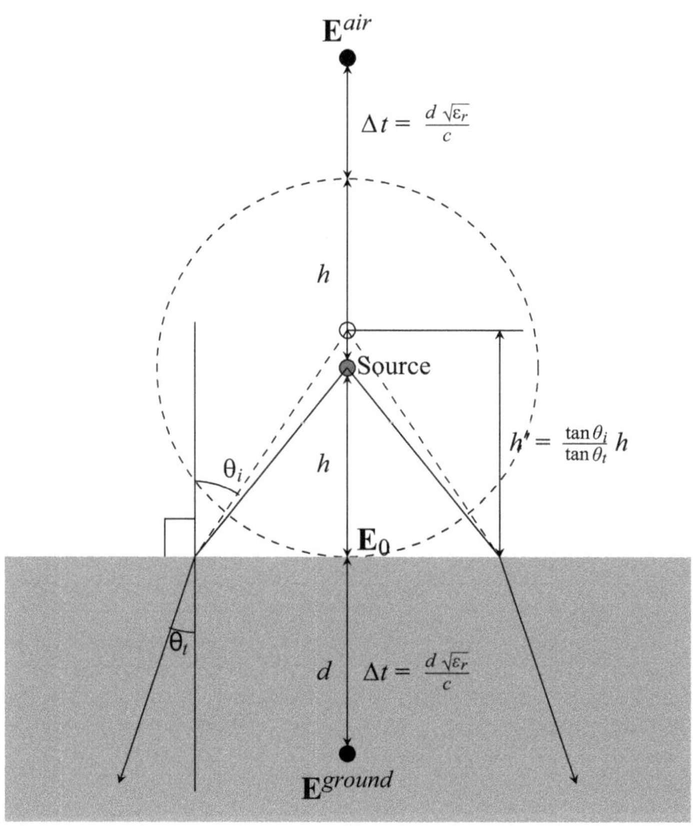

Figure 3.12 Problem description for ground/air field ratio calculation

and the reflected air wave as

$$E^{air}_{reflect} = \Gamma E_0 \frac{h}{h + \left(h + \sqrt{\varepsilon_r} d\right)} e^{-jk_0\left(2h + \sqrt{\varepsilon_r} d\right)} \tag{3.36}$$

The ground-transmitted wave is given by

$$E^{ground} = TE_0 \frac{h'}{h' + d} e^{-jk_0\sqrt{\varepsilon_r}} \tag{3.37}$$

where Γ and T are the reflection and transmission coefficients at the air-dielectric boundary for waves propagating from air into the dielectric.

The previous simple model gives reasonable approximations but more complex models for the prediction of antenna behavior over a given half-space exist. The radiation of dipole antennas over a homogeneous half-space has been studied analytically [10,11], but in more complex environments, the integral equations for the fields produced by these models require numerical methods of solution. Approximated far-field solutions have been alternatively developed, also providing a useful tool for physical understanding of the problem [12,13]. For example, in [14], the antenna is modeled with a plane-wave decomposition into transverse electric (TE) and transverse magnetic (TM) components, and then geometrical optics is used to derive the directive properties of the antenna by using the method of images.

Following the approach in [14], illustrated in Figure 3.13, the far-field components of the electric field intensity radiated by an electric dipole with dipole moment p at height h over a half-space can be expressed in the H and E planes, respectively, as

$$E_{1\theta}(\theta) = \omega^2 \mu_0 k_{12} p \sin\varphi \left|\cos\theta\right| e^{-jk_2 h\sqrt{1-k_{12}^2 \sin^2\theta}} T_{\parallel}(k_{12}) \frac{e^{-jk_1 r}}{4\pi r} \tag{3.38}$$

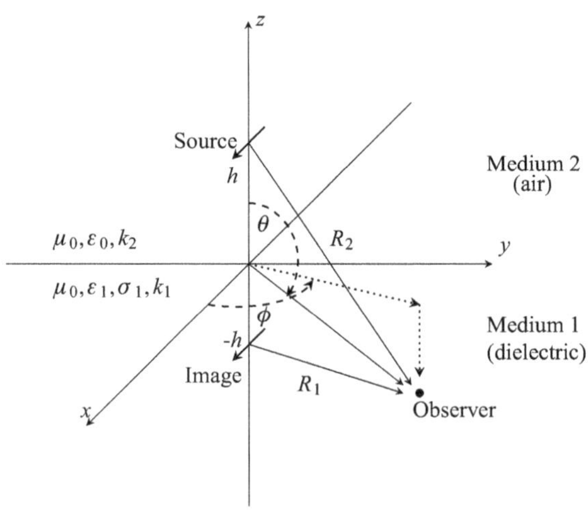

Figure 3.13 Antenna over material interface

$$E_{1\phi}(\theta) = \omega^2 \mu_0 k_{12} p \cos\varphi \, |\cos\theta| \, e^{-jk_2 h\sqrt{1-k_{12}^2 \sin^2\theta}} \cdot T_\perp(k_{12}) \frac{e^{-jk_1 r}}{4\pi r} \sqrt{1 - k_{12}^2 \sin^2\theta} \quad (3.39)$$

where k_1 and k_2 are the wavenumbers in the dielectric and in air, respectively, $k_{12} = 1/k_{21} = k_1/k_2$, $T_\perp(k_{12})$ is the TE dielectric-air transmission coefficient:

$$T_\perp(k_{12}) = \frac{2\sqrt{k_{21}^2 - \sin^2\theta}}{|\cos\theta| + \sqrt{k_{21}^2 - \sin^2\theta}} \quad (3.40)$$

and the corresponding TM transmission coefficient is

$$T_\parallel(k_{12}) = \frac{2k_{21}\sqrt{k_{21}^2 - \sin^2\theta}}{k_{21}^2 \, |\cos\theta| + \sqrt{k_{21}^2 - \sin^2\theta}} \quad (3.41)$$

As an example of results the model can produce, Figure 3.14 shows the radiation pattern of an infinitesimal dipole over a lossless dielectric medium with $\varepsilon_r = 9$ at

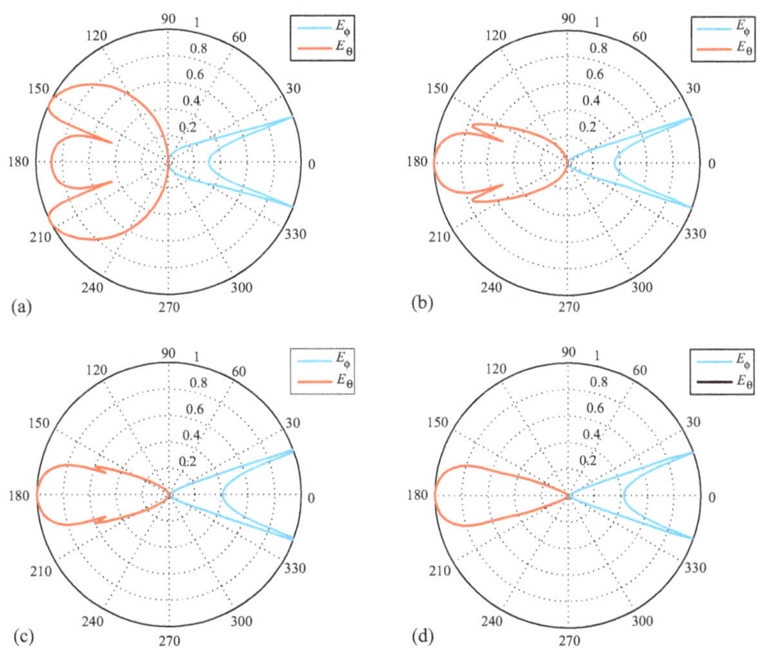

Figure 3.14 Far-field H-plane (E_ϕ) and E-plane (E_θ) patterns for dipoles above the interface between air and a lossless dielectric with $\varepsilon_r = 9$ (a) $h/\lambda_0 = 0$, (b) $h/\lambda_0 = 0.1$, (c) $h/\lambda_0 = 0.2$, and (d) $h/\lambda_0 = 0.35$ (normalized to 1). The dielectric surface is located on the 90°–270° degree axis in all cases

several heights. When the source-interface distance is increased, the antenna field patterns are less modified by a reduction of the effect of the reactive field. Therefore, for this specific geometry and type of antenna, the closer the antenna is to the half-space, the more energy is lost in sidelobes. However, when antennas are far from the host medium propagating losses, clutter, noise, and cross-coupling increase. As a rule, clearer target signatures are obtained for antennas close to the media under test.

As useful as models are, this basic approach cannot suffice for real-word problems, where more detailed descriptions of the scenario are required. To this end, numerical methods to solve the so-called direct problem, that is, the calculation of fields in the host medium, in the presence of targets are often used. The more common and useful methods for this purpose will be described in Chapter 5.

3.4 Antenna types for ground penetrating radar

In general, most of the well-known antenna designs for telecommunications, covering almost any shape and size for the operating range of center frequencies from 10 MHz to 5 GHz, have been explored in literature as potentially useful GPR antennas [15]. However, the peculiarities of GPR scenarios indicated only a small set of them as appropriate. In this sense, both the center frequency and the operating bandwidth of the antenna are the main characteristics to take into account when choosing a specific design. As mentioned before, the selection of the center frequency is usually made in the design stage of the survey, and it involves a compromise between size, resolution, and effective depth of detection of the object under test. On the other hand, the operational bandwidth of the antenna is closely linked to the electronics of the GPR system in use and, in general, the goal is to make it as large as possible and coherent with the feeding pulses provided by the transmitter circuit [16,17]. As well, the efficient coupling of energy from air to the structure under test is a major issue, because the same GPR antenna has to be versatile in the sense that it should operate over soils of different permittivities and conductivities. Last, but not least, the scattering properties of the buried object affects the antenna choice mainly through the polarization of the transmitted/received electromagnetic waves. For antenna arrays, the requisites of the antenna elements are similar to those of monostatic and bistatic GPRs with the additional factor of coupling between antennas, a matter closely related to the RF shielding of antennas to prevent interferences from internal (e.g., amplifiers and power devices) or external sources (e.g., radio, TV towers, and mobile-communication antenna arrays). Given the large number of possible types, it is useful to be able to classify the antennas. GPR antennas can be presented and classified in several ways according to distinct criteria.

A first classification of antennas can be made according to the antenna–soil interaction as (1) borehole antennas and (2) ground-coupled antennas. The first group of antennas are designed to be in contact with the structure under test. For borehole prospection [18,19], the transmitting antenna is located in a borehole and, depending on the location of the receiving antenna, several subtypes of borehole radars can be implemented: single borehole (if both transmitter and receiver antennas are in the same borehole), cross borehole (if a separate bore hole is used for the receiver antenna), vertical borehole (if the receiver antenna is placed on the surface of the ground), and tunnel or mine borehole (in cases in which some existing location exists and can be used to accommodate the receiver, such as a mine shaft or a tunnel). In general, the single and crossed borehole types are constrained to cylindrical holes, so that thin-wire dipoles with cylindrical symmetry, discussed later in this section, tend to be the natural choices for these systems. As long as the required directivity of the dipoles is moderate, some other configurations can be (and have been) implemented: the Adcock antennas [18], whereby an array of properly fed vertical dipoles is closely spaced to create directive patterns, and the crossed magnetic loops [19], which achieve directional radiation by the superposition of the circularly polarized waves emitted by each loop. However, considering the fact that borehole surveys are usually done at low center frequencies (up to a few tens of MHz) and also taking into account the size limitations of the boreholes (diggings and drilling can be difficult in hard soils), antenna borehole solutions with high directivities are not practical. For vertical or mine boreholes, there are more options for receiving antennas, and in those cases ground-coupled geometries can be used to get not only higher signal-to-noise ratios (SNRs) but also decoupled polarizations. For the second group, the ground-coupled antennas, the higher gain allows better resolution in distinguishing small targets or thin layers. As was discussed in the previous section, the constitutive parameters of the soil impact the radiation parameters of the antenna mainly because of the near-field coupling of the first reflection of the radiated field. To reduce this effect, an air–ground calibration is implemented, usually through signal processing techniques (to be discussed in Chapter 4). Some additional actions can help to this end: shielding of the antenna allows for effective spatial filtering by keeping a constant air-to-ground distance; and custom-designed transport mechanisms smooth the impact of surface inhomogeneities (e.g., vegetation or local humidity variations) by attaining a constant speed in survey paths, no matter the size and weight of antennas. However, the coupling between antenna and ground is rarely a perfect match and varying local permittivities and conductivities produce impedance mismatches even for calibrated antennas, thus causing additional signal distortions.

Another way to group GPR antennas is afforded by looking at their operational bandwidth. Even though a wide bandwidth is always a desired feature, the type of modulation of the feeding/sampling pulse provided by the transmitting/receiving electronics imposes a basic constraint: the operational bandwidth of the antenna must be greater than the bandwidth of the transmitting and receiving circuits. For this reason, and considering that the first commercial GPRs worked with frequency modulation, traveling-wave antennas such as the bowtie and broadband dipole antennas were commonly found in earlier commercial equipment. Although the resonances of these antenna configurations is large enough to provide a reasonably wide band of operation (20%–50%), an additional increase in the bandwidth can often be achieved by using lumped loads on the outer ends for the bowtie [20] or by a Wu–King profile in dipoles [21], at a cost of decreasing radiation efficiency. To ease this drawback, a compromise solution can be obtained by adding low-noise power amplifiers to the transmitter and receiver electronics. Other classical traveling-wave antennas, such as log-periodic antennas, are not used in survey equipment because of the large vertical size but can be used in common experimental laboratory facilities such as sandboxes [22]. Even so, in these facilities the highly directive horn antenna is the most common choice for precise measurements [23].

An alternative set of broadband antennas is inspired by the concept of frequency-independent (FI) geometries. Following [24], if the shape of an antenna is completely specified by angles, the impedance and pattern features are practically independent of frequency for all frequencies above a minimum cutoff value. The paradigm of this group is the equiangular spiral antenna, in which a change in frequency only rotates the radiating area along the spiral arms, defining thus the active region of the antenna. As long as the arm length is large enough, any frequency can effectively be radiated according to a scale factor that determines the actual size of the antenna in relation to the frequency of operation. Of course, the limitations in practice for extremely large sizes impose a lower cutoff frequency. However, there is a primary issue preventing FI antennas from being used in GPR systems. The issue is pulse dispersion, arising not only from shifting of the active region as a function of frequency but also from the unavoidable radiation losses of curved geometries [25].

GPR impulse antennas are closely related in the final objective, that is, the maximization of the operational bandwidth, but their design is approached from a time-domain perspective. Considering the fidelity parameter, an ideally constant impedance with linear phase is achieved if the voltage-feed and current in the pulses at the input terminals are identical in waveform. Under these conditions, if the bandwidth of the feeding pulse is large, similar to the impulse/delta signal as for instance in a Gaussian pulse, a large fidelity implies large bandwidth. Antennas designed to maintain fidelity of ultra-wideband (UWB) feeding pulses are called UWB antennas, and these are the most common radiating structures for impulse radars. Some physical insight is required to understand time-domain electromagnetic radiation [6]. Once a differential voltage is applied to the input terminals, positive/negative pulse charges are accelerated in the structure. These pulse charges, and the subsequent current pulses propagate along the geometrical features of the antenna even after the pulse has dissipated at the excitation port. Since partial reflections of the current pulse, an effect called ringing, is produced by discontinuities or curved geometries, an important goal in design is to prevent ringing as much as possible or, more likely, to reduce the amplitudes of ringing signals. Once the antenna is properly calibrated with this information, any ringing detected in surveys with UWB antennas is associated with reflections from discontinuities in the media, such as the ground, or potential targets. However, a true non-ringing antenna is physically impossible for a Gaussian feed as a consequence of Maxwell's equations, which states that it is only possible to couple between time-varying electric and magnetic fields, and frequency-zero components do not radiate energy into space. Therefore, for impulse antennas, the designer usually looks not only to minimize or prevent ringing in the structure but also to optimize the input waveform. Examples and design procedures for UWB antennas, such as the Vivaldi or the Wu–King dipoles, can be found in [26].

A final reminder, before choosing a particular geometry, is to pay attention to the polarization of the radiated waves. In most cases, linearly or circularly polarized antennas are selected taking into account the properties of target scatterers. Nevertheless, this choice is not universal. Linearly polarized loaded dipoles or horn antennas, with optimum radiation parameters in terms of bandwidth, can be inappropriate solutions in some practical applications. For instance, an optimum UWB antenna choice for impulse radar applications is the hexagonal horn with abrupt radiator, which operates from near 0 to above 12 GHz [5]. It has inherently constant group delay versus frequency due to the abrupt radiator design, which causes all frequencies to radiate from the same phase-center, located on the launcher plate. Additionally, it has very low cross-coupling due to the horn-waveguide configuration. However, this antenna is linearly polarized and it has manufacturing issues that limit its use in portable radar applications. The horn structure must be soldered and manually tuned due to the complicated geometry of the launcher plate, inset within the horn wall. This raises costs while reducing repeatability and consistency in the manufacturing stage. An alternative circularly polarized solution is the FI spiral antenna that, in addition, is less expensive because it can be fabricated with affordable micro-strip equipment.

As a reference, Tables 3.1 and 3.2 show a number of GPR antennas designed for different applications. These tables include some of the typical GPR antennas such as bowtie, Vivaldi, horn, loaded dipoles, spiral and tapered slot antennas. The tables also give additional information, including coupling method and bandwidth. More technical and methodological aspects concerning the design of antennas for GPR applications can be found in [27–29].

Table 3.1 Antennas suitable for applications using time-domain waveforms

Reference	Application	Antenna	Coupling	Bandwidth
[30]	Forest litter	Horn	Air	0.8–5.2
[31]	Pavement inspection	Half-ellipse	Air	0.25–0.75
[32]	Landmine detection	Bowtie	Air	1.5
[33]	Forensic survey	Horn	Air	0.27–0.9
[34]	Boundary layer detection	Loaded dipole	Ground	0.25–0.75
[35]	Through the wall detection	Horn and vivaldi	Air	1.45–4.5

Table 3.2 Antennas for applications using frequency-domain waveforms

Reference	Application	Antenna	Coupling	Bandwidth
[36]	Pavement inspection	Aperture array	Air	0.2–3
[37]	Generic	MIMO array	Air	0.3–4
[38]	Generic	Horn	Air	0.8–4
[39]	Dispersive media	Vivaldi	Air	0.8–3
[40]	Generic	Bowtie	Ground	0.2–0.6
[30]	Soil hydraulic	Spiral, Vivaldi, and bowtie	Air	0.8–5.2
[41]	Layered media	Horn	Air	0.8–2
[42]	Soil permittivity	Loaded dipole	Ground	0.5–4.5

3.4.1 Dipole antennas

In spite of being the simplest antennas, dipoles are frequently used for GPR because of their simplicity, low cost, portability, low weight, and typically small size. The dipoles are linearly polarized, have low directivity, an omnidirectional radiation pattern, and a narrow bandwidth associated with the inner resonances at the end of the arms. To improve performance, there have been a number of developments and modifications to the basic dipole, usually targeting specific areas of improvement. In most GPR applications, the drawback associated with resonance is mitigated through the use of passive loading [43]. Resistive loadings are introduced to reduce the undesired reflections that occur along the antenna as well as the associated distortion of the radiated signal, thus improving substantially the bandwidth at the cost of a lower efficiency [21]. Reactive loads are used to tune the operational bandwidth of the dipole [44].

Another technique specifically useful in improving the matching of dipoles is to bend the wire antenna into the form of a slender rectangular loop with the ends folded back around and connected to each other. This antenna is known as the folded wire dipole or simply folded dipole (Figure 3.15). A different design was proposed [45] to solve the problems related to the low directivity of dipoles, a property that affects the probability of detection because of low SNR. The loaded Vee-dipole (Figure 3.16) folds in the arms of the dipole to create directional patterns. The exact performance depends on the dimensions and angles, but in general the antenna has been proven effective in the detection of antipersonnel land-mines, where failure rates in surveys should be, ideally, zero. The antenna radiates pulses with a rich frequency content leading to improvement in resolution. Limitations in detection due to the inherent linear polarization of dipoles can be solved by using a crossed dipole antenna as shown in Figure 3.17. This configuration is sensitive to targets other than planar interfaces parallel to the plane of the antennas. Finally, for low-frequency applications where depth rather than resolution is a priority, such as the search for large underground structures, including caves, tunnels, and buried bunkers, resonant dipoles are rolled

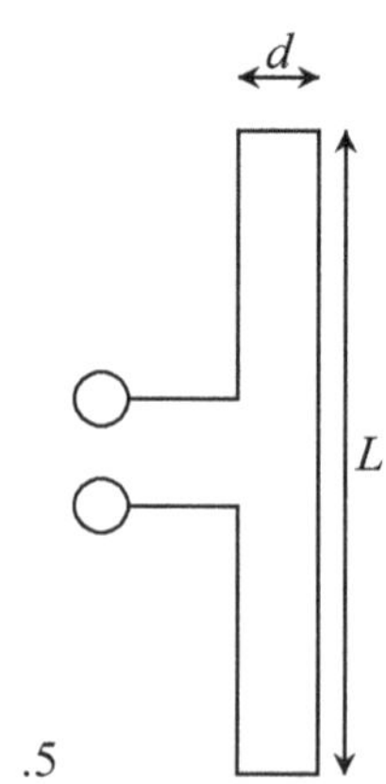

Figure 3.15 Folded dipole

to operate at a range of resonances [46]. The rolled dipole (Figure 3.18) reduces the horizontal length of the antenna considerably, making it suitable for GPR equipment trolleys. In addition, resistive loading with Wu–King profiles can also be added for ringing suppression.

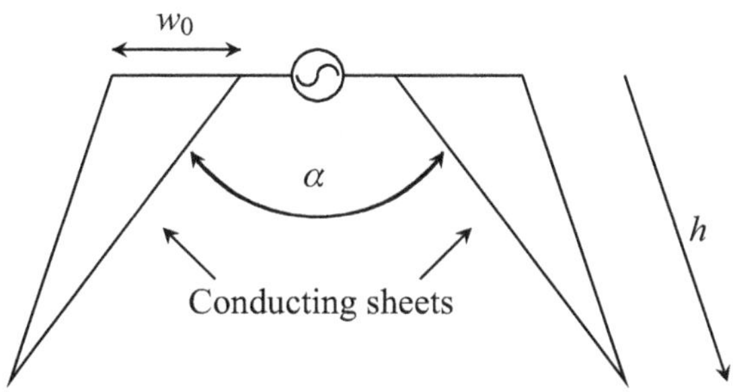

Figure 3.16 Resistively loaded Vee dipole

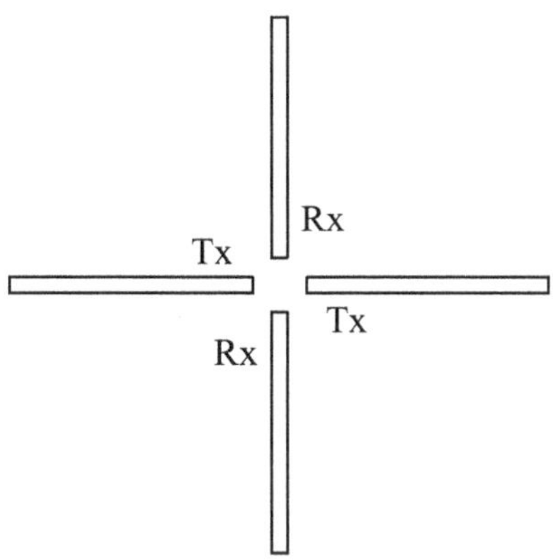

Figure 3.17 Crossed-dipole GPR antenna

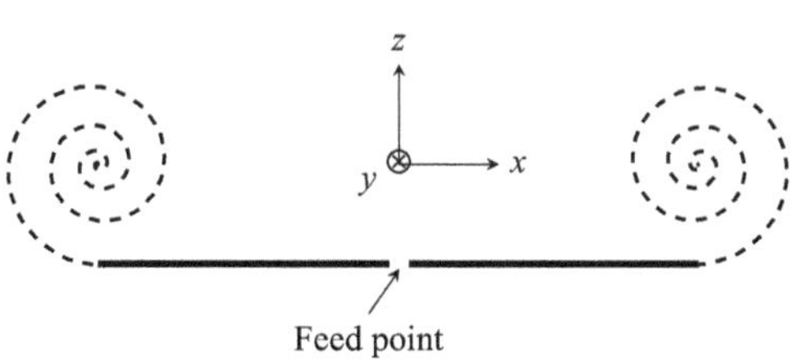

Figure 3.18 Rolled dipole antenna

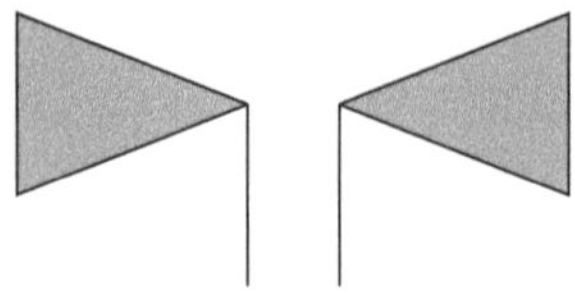

Figure 3.19 Bowtie antenna

3.4.2 Bowtie antennas

Bowtie antennas are close to dipoles in terms of frequency range. Because they are low profile and low weight, with reasonable performance, bowtie antennas are often used in commercial GPR systems. The bowtie antenna (Figure 3.19) is a planar version of the finite biconical antenna, which is considered the canonical FI antenna. In a way, the geometrical resemblance with both FI antennas and finite length dipoles explains the main properties of bowtie antennas: not only they resonate as dipoles, and thus they have limited operational bandwidth, but also their radiation parameters remain uniform given a change in scale as in FI antennas. This is critically important in small, portable GPRs where volume is at a premium. In addition, they are simple, rugged, and easy to produce. For these reasons, bowtie antennas have been successfully applied in different scenarios regardless of the constitutive parameters of both host and target media. Since the currents in a bowtie antenna are abruptly terminated at the ends of the fins, similarly to dipoles, ringing pulses do appear and contribute to clutter signals. However, bowtie antennas have wider bandwidths compared to dipoles (typically a 2:1 bandwidth), which can be substantially improved by resistive loading. In terms of directivity, bowties have narrower main-lobe BWs with the main lobe perpendicular to the plane of the antenna. Directivity of bowtie antennas is rather low but can be improved considerably through optimal adjustment of length and flare angle. Although there are better antennas for specific applications, the bowtie antenna is often seen as an overall good compromise for most applications.

Bowtie antennas are usually situated in an air-to-ground configuration, close to the structure under test, and therefore, radiation properties can be affected by the electrical parameters of the host structure. For this reason, the footprint of bowtie antennas has to be considered to assess the amount of effective power propagating to potential targets. To help one with this task, some modifications of the original bowtie can be resorted to. To avoid interferences from other systems, the bowtie can be surrounded with a conducting shield (Figure 3.20). The shield increases the power radiated into the host and alters the footprint in comparison to a free-space bowtie. Tunable bowties can be obtained by using wire bowtie antennas (Figure 3.21). This design reduces ringing by means of resistive loads in the antenna's arms. These wire bowtie antennas also have better coupling to the ground and improvements in terms of the footprint [47,48]. As a final example, planar bowtie antennas can be printed on a substrate (Figure 3.22) and fed by a coaxial line, resulting in excellent sensors with small dimensions, low weight, and low power losses.

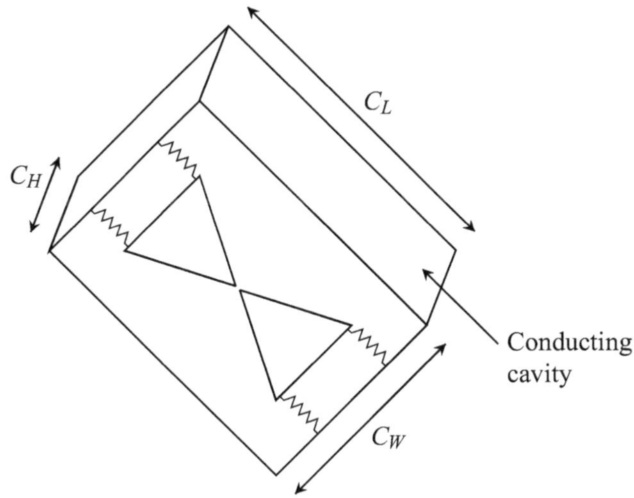

Figure 3.20 Shielded bowtie antenna

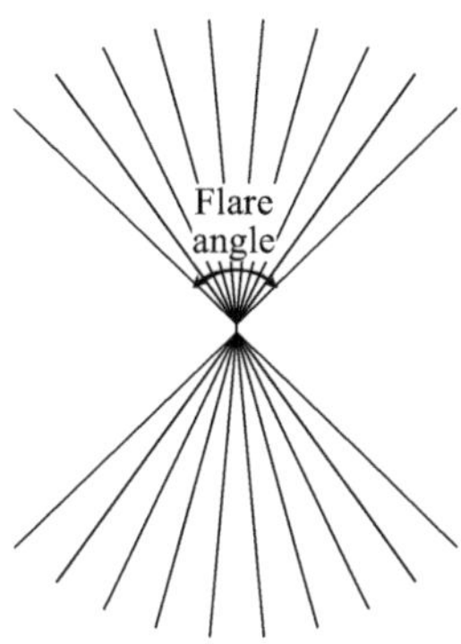

Figure 3.21 Wire bowtie antenna

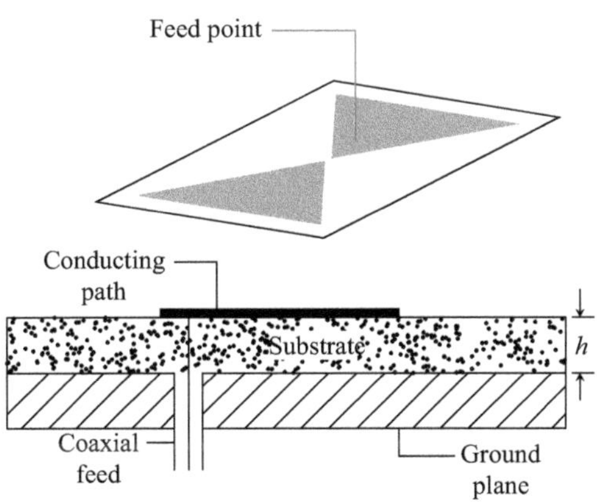

Figure 3.22 Printed bowtie antenna

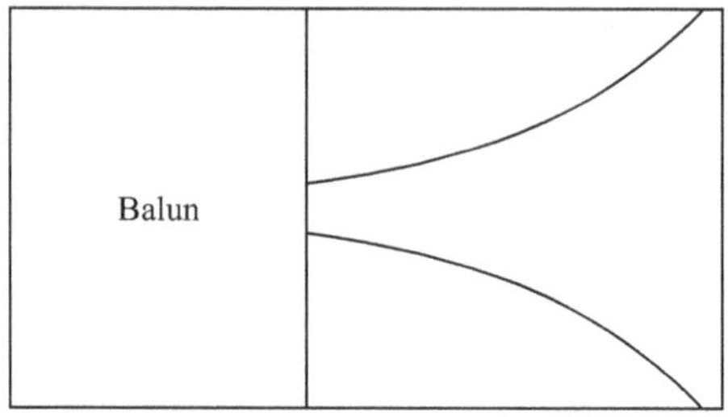

Figure 3.23 Vivaldi antenna connected to a planar balun

3.4.3 Vivaldi antennas

When it comes to wideband traveling wave antennas, the most common antenna in use is the Vivaldi antenna. Unlike resonant or standing wave antennas, traveling wave antennas rely on traveling waves propagating in their structure from the feed point, to the unbounded medium. The antenna is typically fed through a waveguide. The transition between the waveguide and the antenna must be matched to minimize losses and maximize the bandwidth of operation. The radiation patterns are usually those of end-fire antennas with BWs approximately equal in both E- and H-planes. The main beam has significant gain with relatively low sidelobe levels and both are constant over a large frequency range. The Vivaldi antenna is based on a geometry with continuous scaling and gradual curvature of the radiating structure, following an exponential function (Figure 3.23).

Vivaldi antennas are typically built on a high permittivity substrate [49], which facilitates the integration with microwave and millimeter wave circuits. To a considerable degree, this type of antenna can be used in applications that require embedded electronics of the type used in communication systems in many fields, including military, aerospace, and remote sensing applications. In addition, Vivaldi antennas can be conformal since dielectric substrates can be bent (depending of course on their thickness). Figure 3.23 shows a schematic representation of the Vivaldi antenna connected to a planar balun, the latter being used to match the antenna to its waveguide.

3.4.4 Spiral antennas

Spiral antennas are a particular class of FI antennas with circular polarization, used for both pulsed and stepped frequency GPRs [50]. They not only share the common performance of FI antennas in terms of a constant radiation parameters over a large bandwidth but also have the same drawbacks, that is, their practical implementation is limited at the upper and lower frequencies because of the truncation of the structure to a finite size. The smallest geometrical element of the antenna defines the lowest wavelength of operation, whereas the largest antenna dimension is of the order of one-half of the maximum wavelength of operation [51]. The equation that generates the spiral curve in polar coordinates is determined uniquely by angles, as expected for any FI antenna. The width of the arms increases logarithmically to keep the

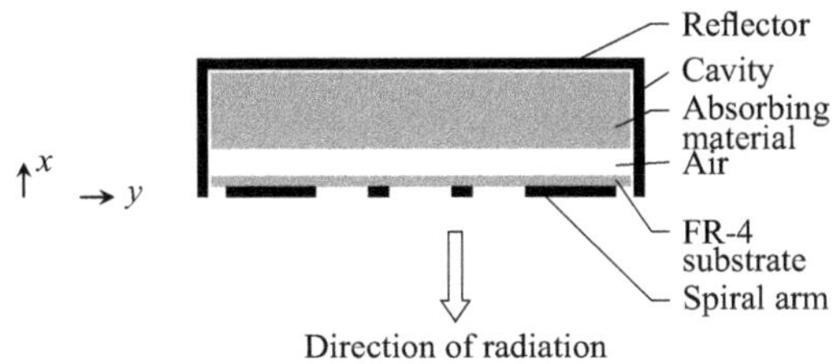

Figure 3.24 Spiral antenna shown in cross-section

electrical size constant as the frequency of operation and the active region shifts. The most common implementation is the planar equiangular spiral antenna, which possesses circular polarization. Circularly polarized radiated waves have important properties in terms of sensing of linear buried targets but, at the same time, this type of geometry also produces unavoidable phase dispersion [5], an effect that hinders the identification of shallow clutter by expanding the tail of the radiated impulse signal enough to hide scattered pulses. There are other versions of these antennas, such as spiral slot antennas, made by creating an aperture in the shape of the spiral in an otherwise conducting plane, whereas the classical spiral antenna is made by placing (actually often by etching) the conducting spirals on a substrate. Figure 3.24 shows a cavity-backed spiral antenna in cross-section through the spirals. In this particular implementation, the antenna is etched on a printed circuit board. A log-spiral antenna is shown in Figure 1.4 and two types of spiral antennas are depicted in Figures 3.36 and 3.38.

3.4.5 Horn antennas

In communications, horn antennas are very attractive devices in terms of a high directivity, low dispersion, low reflection coefficient, easiness of integration with waveguides, and the availability of closed-form equations for design. Despite these exceptional features, their role in GPRs is somewhat limited and, although they have been used in short-pulse radar and with frequency-modulated continuous wave (FMCW) radar, bowtie antennas are still preferred in commercial equipment primarily because they are planar. Reasons for this preference are not only the better air–ground coupling discussed earlier (Section 3.4.2) but also because of the weight (up to 8 kg), size and geometry of horn antennas, which complicate the handling and assembly in field survey equipment. Also, the larger aperture of horns makes them prone to external interferences. Since shielding horn antennas is a difficult task, digital processing techniques are used instead to separate the RF interference from the target readings, by means of RF filters integrated with the receiving antenna. Additionally, real-time signal processing algorithms can be implemented to calibrate the signal received by the antenna.

The pyramidal horn geometry is illustrated in Figure 3.25, shown in cross-sections. The waveguide has broad and narrow dimensions a and b, respectively, and

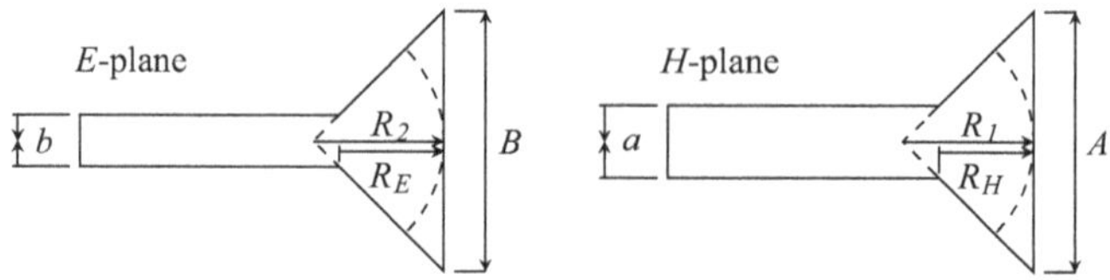

Figure 3.25 Pyramidal horn antenna shown in two views

the corresponding dimensions of the horn aperture are A and B. The apex-to-aperture axial horn length is marked as R_1 in the H-plane and R_2 in the E-plane, and the waveguide-to-aperture lengths are R_H and R_E. For a pyramidal horn, $R_E = R_H$ [52].

Horn antennas have been proven effective for inspections of highway pavements and bridge decks. In these applications, air launching from antennas mounted on a vehicle can be used and large dimensions and weight can be tolerated. In this type of antennas, a calibration is required for them to work properly. To this end, it is sufficient to apply a time-windowing procedure, based on previous measurements collected from varying-height readings of a metal plane on the surface of the host medium. The attachment of the horn antenna to the vehicle must be separated from the vehicle body by at least 1 m to avoid interferences. Impressive spatial resolution up to 1 cm in scannings at speeds upward of 90 km/h have been reported. Apart from this type of application, horn antennas are still the preferred kind of antennas for laboratory facilities, where complete control of the electromagnetic properties of the media provides accurate results. On the other hand, horn antennas are entirely unsuitable for portable field equipment. Considerable efforts have been devoted to the improvement of horn antennas for GPR resulting in modifications of the basic horn to accommodate specific needs. An example is the exponentially tapered TEM horn antenna loaded with absorbers to produce wideband characteristics and reduce reflections [53]. Another design is focused on achieving improved impedance matching through structural optimization of the horn [54]. The dielectric wedge antenna, which is a modification of the TEM horn, is based on tapering the antenna metal flairs to achieve better coupling to the ground [55]. Wideband horn antennas can provide constant high gains over a 10:1 bandwidth by using a logarithmic geometry for the ridges in what are called ridged horn antennas.

Horn antennas are some of the best available and their use in communication promises continuous improvements and adaptations, some of which will undoubtedly find their way in future GPR applications.

3.4.6 Antenna arrays

So far the discussion focused on the use of single-element antennas. There is, however, another possibility—that of using multiple-element antennas or multiple antennas to achieve some of the properties required for a successful system. Array antennas are common in communication systems and have been adapted for use in GPR as well. The properties of antenna arrays, defined as an ensemble of single-element antennas (either the same type or different types) in a variety of configuration, radiating simultaneously, can be used to enhance GPR surveys. Array antennas have properties that cannot easily be attained with single elements. An array can radiate more power, may have higher gain, and can be more directive. Shaping of the radiation pattern to suit specific needs is one of the attractive features of arrays. In that respect, it is also possible to steer the radiated field in a desired direction by changing the phase, at the feed point, of the various elements in the array in what is commonly called a phased array. Steering can be fixed, that is, one can steer the radiated field onto a fixed direction or it can be adaptive, based on given requirements such as scanning in a desired pattern. As a rule, it is possible to produce almost any desired radiation pattern by adjusting both amplitudes and phases, usually determined by optimization techniques [56], in both linear arrays (Figure 3.26(a)) (in which elements are arranged on a line), or 2D arrays (Figure 3.26(b)) in which the elements cover an area over which they are distributed in some fashion. Other recent approaches, known as digital beamforming or scanning techniques, are aimed at controlling the main lobe of the radiation pattern [57]. Prior to this, beamforming was done by physically varying the location and/or orientation of individual elements. In modern approaches, a phased array of fixed antennas controlled electronically can reproduce this effect faster than with conventional mechanical arrays (a term referred to as agility). Another type of array is the adaptive array, which consists of a number of antennas, properties of which can vary in response to changing requirements in terms of the required power and the direction of maximum gain.

In GPR, arrays of antennas are used to reduce clutter, specially that caused by reflections from the ground, as well as to increase the resolution through the larger directive patterns and narrow bandwidths produced by arrays. For instance,

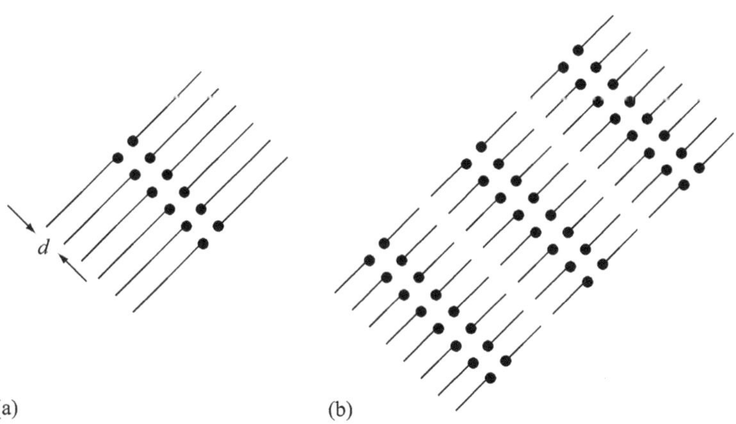

Figure 3.26 (a) A linear array of dipole antennas and (b) a two-dimensional array of dipole antennas

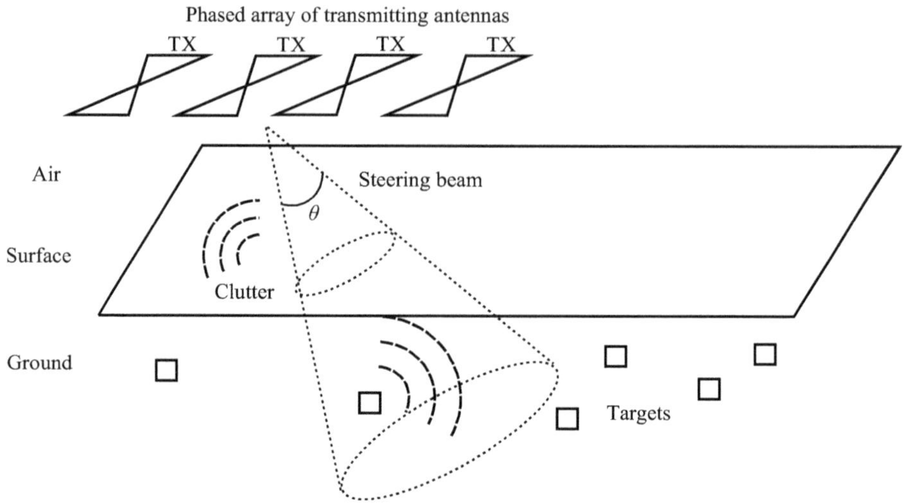

Figure 3.27 Array of transmitting antennas. Adapted from [58]

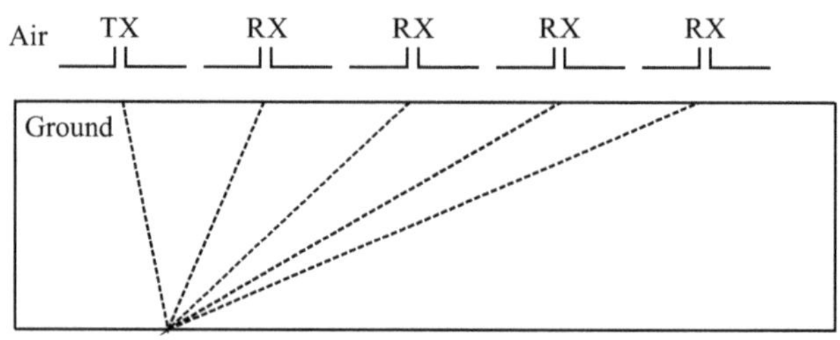

Figure 3.28 Array of receiving antennas

Figure 3.27 illustrates an array with separate elements for which the beam position θ is given by the following relation [58]:

$$\theta = \sin^{-1}\frac{\gamma\lambda}{2\pi a} \tag{3.42}$$

where γ is the incremental phase shift (i.e., the phase difference between two consecutive elements in the array) and a is the element spacing (i.e., the distance between two consecutive elements, measured from feed to feed in wavelengths).

The array can be used to improve readings not only as a transmitting array but also as a receiving array (Figure 3.28). However, it is important to note that the use of antenna arrays for GPR has significant drawbacks in terms of the size of antennas (i.e., as a consequence, arrays are usually mounted on the front of a vehicle) and the complexity of the transmitting/receiving electronics. Mutual coupling between the elements and indirect coupling with reflectors in the environment have to be accounted for prior to the use of this type of antenna. For this reason, post-processing techniques in the form of digital filtering and noise removal are more critical for arrays than for single-element antennas.

3.5 Antenna design for GPR systems

Every GPR application imposes different limitations and challenges, and therefore different types of antennas are required for each class of application. For instance, some applications require high depth of penetration with low resolution, whereas other applications require other characteristics. The trade-off between resolution and depth of penetration is addressed when the antenna central frequency is chosen thus defining the bandwidth and other relevant factors related to frequency and bandwidth. On the other hand, target identification usually imposes stricter requirements on antennas in comparison with a system for merely target detection, because it adds post-processing tasks based on pattern recognition algorithms, the rate of success of which is related to the level of details of the scattered signal.

It is also important to notice that any antenna can be thought of as an independent subsystem, with its own transfer function that can be frequency dependent, with linear or nonlinear phase response. The latter is a major consideration in analysis of the response from impulsive GPRs where phase distortion may be a serious issue. On the other hand, in frequency-modulated or synthesized GPRs, the requirement for linear phase response from the antenna can be relaxed, allowing more complex designs, because this effect can be corrected through system calibration.

From an engineering point of view, the most relevant features for GPR antennas are as follows:

- The antenna should have a wide frequency bandwidth for better resolution. The maximum depth for successful detection decreases when frequency increases, with an upper frequency limit of about 5 GHz for most subsurface GPRs. Some examples are shown in Tables 3.1 and 3.2 that also list the waveform and the coupling to the structure under test.
- The antenna should keep performance constant across its operational band, including the radiation pattern, gain, impedance matching, and level of dispersion. In particular, high gain and narrow bandwidths are fundamental to resolve targets in close proximity.

From a practical point of view, most applications require the antenna to be small and light weight. Compactness and rugged physical characteristics are especially important for mounting of the system in confined spaces, a consideration important to the ease of motion in surveys. However, these requirements are not fully independent of the electrical parameters. An obvious example is antenna length which, in turn, is related to the central operating frequency and, therefore, to the power radiated [5].

Antenna design procedures are first and foremost related to the operating domain and pulse modulation of the GPR system. Although most of the commercial GPR systems use time-domain waveforms, considerable research efforts have been expended on the development of frequency-domain systems and appropriate antennas for these as is shown in Table 3.2. The frequency-domain approach has advantages over the time-domain systems in terms of power requirements and provides more information about the target characteristics. Table 3.2 shows some recent efforts for some applications that use frequency-domain waveforms, in which imaging is compiled in terms of magnitude and phase in contrast to the time-domain systems, which restricts analysis to magnitudes. As a consequence of the characteristic phase shift inherent in the reflection of electromagnetic waves, phase-dependent features in frequency-domain systems are important for the interpretation and prediction of buried targets.

Table 3.1 summarizes recent work on various GPR applications for time-domain systems. Time-domain systems are basically characterized by sending a pulse or impulsive waveform into the structure under test. This type of waveform has a short time period and, if late-time ringing is to be avoided, a large bandwidth. As was mentioned earlier in this chapter, late-time ringing can be produced not only by discontinuities of the current in the antenna, as, for example, in bowtie and dipole antennas, but also by external artifacts such as the coupling to the structure under test. Late-time ringing by external sources often blurs target information, and it is not always possible to filter out these effects by post-processing techniques. For instance, the air–ground reflection of radiated waveforms strongly depends on the height of antennas over the ground, which changes at different locations of surveys over irregular surfaces, and horizontal filtering is ineffective in those cases.

In summary, radiated waveforms are probably the major parameter of interest in GPR antenna design. Another important feature according to Tables 3.1–3.3 is the antenna input impedance, which is also affected by the distance between the antenna and the structure under test. Additionally, the behavior of the input impedance affects the transmitting and receiving electronics, which become simpler for constant input impedances and therefore simplify the hardware of the system. Other factors indirectly related to the performance of the antenna are the weight, size, power consumption, complexity and off the shelf availability of hardware, and finally its cost. Spectral regulations in each country can also play a role in choosing the operational frequency and bandwidth, affecting thus the achievable resolution and depth of penetration and ultimately acceptable antenna designs for GPRs.

3.5.1 GPR system parameters

Interestingly, manufacturers of most of the GPR systems offer a number of configurations achievable with available optional system components. As long as the system configuration directly affects the performance of the NDT assessment, any prior information about the structure under test and the target characteristics has to be accounted for in any configuration, all within the available options. In order to understand the needs and options for tailoring antennas in each case, GPR system parameters are discussed next. Although a fuller discussion of system parameters and performance will be undertaken in the following chapter, elements of system parameters that are directly associated with antenna design and performance are introduced here briefly.

3.5.1.1 Waveforms

As was briefly discussed in Chapter 1, one of the more obvious but also difficult issues in radar is the waveform transmitted and the resulting received signal. Considerable work has been done in defining appropriate waveforms and their effect on performance in both time- and frequency-domain radars. The choice of waveforms, including modulation methods, affects all aspects of GPR design and performance from its internal electronics, through antennas and choices in signal processing, signal interpretation, and generation of images. Figure 1.6 shows the most relevant choices for time- and frequency-domain GPR systems [59,60]. GPR systems that shape a time-limited waveform and detect changes in its amplitude due to scatterers are known as impulse GPRs. In pulsed GPRs, amplitude modulation of a continuous wave (CW) is achieved by applying amplitudes 1 and 0, that is, turning the power on and off at varying times during the inspection. Since the transmission and reception times are well defined in measurements, it is possible to associate the scattered signals with the range of the target.

On the other hand, frequency-domain or CW GPRs are based on band-limited signals. This type of GPR transmits a signal of theoretically infinite duration, as a continuous sine wave, and receives it simultaneously. In this configuration, it is possible to detect buried targets but it is not possible to resolve ranges since the signals do not change in time. However, frequency-domain GPRs allow control of the power spectral density of the radiated waveforms and, in turn, of the resolution of the survey. A comparison between impulse and CW GPR can be found in [61, 62]. The impulsive system is still more common than CW options. However, the development of digital signal processing technology, resulting in cheaper, faster, and more accurate equipment, as well the development of inverse algorithms for GPR [30,63,64] is making CW systems a growing interest choice. In this context, advantages and disadvantages of different waveforms for GPR are always dependent on operational requirements. One of the challenges when using impulse waveform is the possibility of producing a pulse short enough to achieve the desired bandwidth, i.e., with suitable fast rise and fall times. The antenna should be designed to avoid ringing effects or distortion of the pulse shape. Another challenge arises if the received pulses are sampled in real time at the analogue-to-digital converter (ADC), since the ADCs must work at sufficiently high frequencies to correctly digitize the waveforms. This comes at elevated costs.

To improve the detection in frequency-domain systems, the signal bandwidth can be widened by employing some form of modulation. By increasing/decreasing the oscillation in the waveform, the FMCW [65] is obtained, in which the differences in frequency are the relevant parameters for detection. This type of GPR system suffers from interference issues. A different method can be employed if the transmitted frequency of tones or pulses shifts in a given interval or steps across a defined bandwidth. This technique is called stepped frequency continuous wave (SFCW) [66]. The frequency step avoids phase ambiguity by measuring the phase difference between returned signals at each frequency. In addition, the hardware associated with SFCW is simpler than that employed in FMCW. A variation of the

SFCW technique involves gating the timing of the transmitter and receiver circuits. As has been commented, a common problem for active radar systems for CW is the strong backscattered signal from the air–ground interface. This undesired signal can overshadow the reflections from actual targets, especially those with low radar cross-section such as human beings, and limit the dynamic range at the receiver, which could be saturated and blocked. Techniques have been developed to address this problem, including the noise-modulated and the frequency-modulated interrupted continuous wave (FMICW) forms, methods that can be used as alternative techniques or combined with the existing ones. FMICW signals have additional advantages over normal FMCW, such as range-dependent received power according to the mean received signal of the gating sequence, which decreases the presence of the "blind range," risk of aliasing, and overall reduction of the received power [62].

3.5.1.2 System bandwidth

Bandwidth describes the range of frequencies over which the antenna can maintain desirable parameters with minimal deviation. In GPR systems, the bandwidth is one of the determining parameters used to decide upon the antenna type. Some antenna types have very narrow bandwidths and cannot be used to transmit a pulses with large frequency content (narrow pulses). Bandwidth is typically quoted in terms of the input impedance. This is due to the reflected waves that are generated with the change in the antenna impedance which, in turn, increases the VSWR. VSWR is directly related to the S_{11} parameter (reflection coefficient) [5].

3.5.1.3 Antenna position

The specification of how close to the surface of structures under test the antenna can operate is another important parameter, because the antenna input impedance is directly affected by the distance between the antenna and the structure (and, in fact, any medium such as shields or proximity of structures). In this sense, the higher above the surface the antenna is, the lower the variation in the input impedance but, similarly to ground-coupled GPRs, lower heights result in clearer received data and greater survey depths because more energy is coupled into the ground. A commitment on antenna position has to be made because surface coupling gives rise to issues of late-time ringing, increasing the difficulties associated with extraction of information necessary for subsequent post-processing techniques. As a reference, a comparison between air-launched and ground-coupled systems can be found in [67] and [68]. See also Section 3.3 for the effect of antenna position on the radiation pattern.

3.5.1.4 Legislation and standards

In the design of any antenna but in particular in the design of antennas for GPR, it is essential to comply with legislation and standards. The IEEE Standard Definitions of Terms for Antennas [1] establishes definitions for antennas and for systems that incorporate antennas as a component of the system. The antenna requirements for GPR systems is defined by the USA Code of Federal Regulations: Technical Requirements for GPRs and Wall Imaging Systems [69]; and technical requirements applicable to all UWB devices [70]. Among other specifications, the Federal Communications

Commission restricts the operation of GPR to frequencies below 10.6 GHz. Other restrictions concern electromagnetic compatibility issues with particular attentions to the effective (or equivalent) isotropic radiated power limits per bandwidth. For example, the limit in the 960–1,610 MHz is −65.3 dBm and in the 3.1–10.6 GHz the limit is −41.3 dBm [70,5]. There are other specifications to be followed, and other countries have different requirements and limits. Although these issues affect mostly the manufacturers of equipment, the user also needs to be aware of them.

3.5.2 GPR antenna optimization framework

To improve the GPR system response, a given antenna topology can be reshaped or modified using a general optimization framework as depicted in Figure 3.29. It consists of an electromagnetic model responsible for the generation of solutions used as training data, an optimization algorithm that uses these data and a user interface that defines the parameters and limits on the optimization process. The electromagnetic model, the reference antenna topology, and the optimization tool can be chosen on the basis of the GPR system requirements as described in the following sections.

3.5.2.1 The electromagnetic model

For the electromagnetic computational analysis of antennas and the entire GPR surveys, the most widely method in use is the finite-difference time-domain (FDTD) method [71,72] (the method will be described in detail in Chapter 5). The FDTD model can cover a wide frequency range and treat nonlinear material properties, making it particularly well suited for GPR applications in the time-domain. In the frequency-domain, the choice is often the method of moments (MoM) [73,74] or the finite element method (FEM) [75,76]. Analytical approaches based on simplified configurations can be used as well. Alternative numerical methods, including open-source codes such as gprMAX 2D/3D [77,78] and MEEP [79], and commercial software such as ANSYS HFSS [80], Altair FEKO [81], CST Microwave Studio [82],

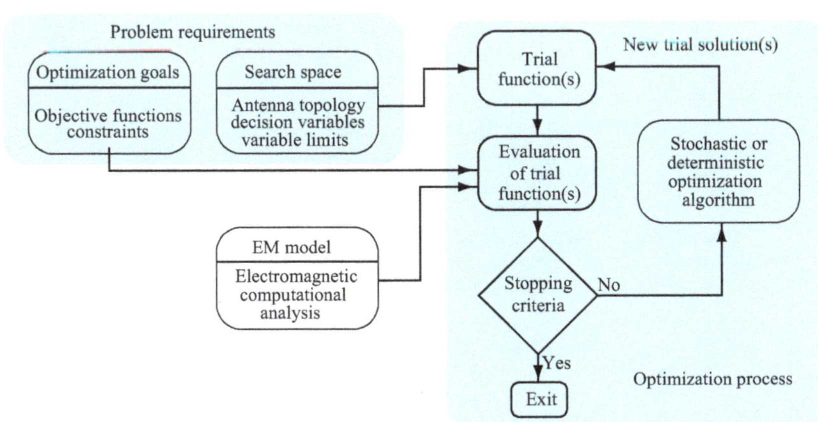

Figure 3.29 Generalized antenna optimization process. Adapted from Fig. 1 in [84]

and GSSI RADAN [83], are often used for this purpose, each with its capabilities, limitations, and associated issues such as cost. In the end, the purpose of any of these is to obtain an electromagnetic model that can generate survey data, with sufficient detail and complexity to be useful. The purpose of the model is to generate the set of trial solutions over which the optimization can take place.

3.5.2.2 GPR antenna optimization goals

Technical and methodological aspects of antennas in GPR applications are presented by [27–29]. Table 3.3 summarizes a selection of relevant works on the design of GPR

Table 3.3 GPR antennas and optimization techniques

Reference	Antenna type	Goals
[85]	Dipole	To fit into a fixed volume, wire antennas made of identical loaded dipoles are used; a resistive profile along the monopoles needs to be chosen to ensure proper behavior of the antenna once they are deployed on the surface. No optimization method is included.
[86]	Wire Bowtie	Antenna geometry optimization using a multi-objective genetic algorithm looking for (1) S_{11} below -10 dBi; (2) maximum gain; and (3) the width of the band where the gain is between its maximum value and 3 dB below it.
[87]	Bowtie	Antenna geometry is tuned for maximum gain and minimum ringing.
[54]	Horn	Structure is designed for impedance matching throughout an ultra-wideband.
[88]	Bowtie	Sharp corners are rounded to minimize the end-fire reflections.
[40]	Fractal Horn	The geometry of a patch antenna is changed in order to satisfy the condition of an S_{11} below -10 dB. An ordinary kriging predictor is used as a surrogate model of the cost function, whereas a differential evolution algorithm is used to effectively minimize it.
[89]	Fractal Bowtie	To reduce the center operating frequency of the miniaturized antenna, a design strategy is given to determine a reasonable distribution of conductive material within a given domain. A gradient-based topology optimization method is applied.
[90]	Spiral	The geometry is an antenna combining equiangular spiral and smooth Archimedean spiral intended for better VSWR performance, axial ratio and gain.
[91]	Horn and Spiral	Optimization of both types of antennas in terms of group delay, axial ratio, and pulse fidelity.
[92]	Vivaldi	Antenna dimensions are optimized in terms of bandwidth, radiation efficiency (S-parameters) and gain.
[93]	Horn	Reflections and ringing effects are reduced by an antenna geometry adjustment.
[94]	Wire Bowtie	VSWR and bandwidth optimization by tapering the arms and applying a genetic algorithm to choose the parameters for the tapered configuration.

antennas and the primary parameters they address. In these works, GPR antenna design focuses on the following three aspects:

1. low cost, accomplished optimally through the use of micro-strip antennas;
2. small size, a goal typically favors the use of planar antennas with small volumes; and
3. high electromagnetic performance, which implies a constant, high gain, high directivity, adequate half power beamwidth (HPBW), proper impedance matching with low VSWR (S_{11}), linear group delay (S_{21}), and high pulse fidelity.

The ability of a directive antenna to effectively radiate a short pulse in a specific direction is fully characterized through return loss, group delay, and pulse fidelity. Once the direction of maximum gain is ascertained, the result for return loss is used to evaluate the frequency range of the antenna; the group delay variation to quantify the pulse dispersion, and the pulse fidelity to assess the differences between input and output pulse shapes in impulse GPR systems.

3.6 The optimization process

Electromagnetic optimization problems are almost always computationally expensive, often nonlinear in nature, and are riddled with conflicting goals [71,72]. There are a number of methods appropriate for this task, some simple, focusing on a particular objective, while others are capable of multi-objective optimization. Evolutionary optimization procedures [95,96] are based on concepts borrowed from biology and are particularly useful. They are robust, stochastic-based methods that can handle the common features of electromagnetic problems. Some methods are based on behavior of swarms, others on concepts of artificial intelligence, and still others on the evolution of living things. Perhaps the best suited for antenna optimization is the multi-objective genetic algorithm (MGA). Like any multi-criteria algorithm, the MGA must generate a meaningful set of samples from the group of all efficient solutions available, called the Pareto optimal front, which represents the trade-offs among the various merit functions for the objectives. The decision maker then chooses one optimal solution from this set. Before looking into the optimization of antennas proper, it is useful to briefly describe the multi-objective optimization algorithm. It is described in the next section as the method of choice in this work, but it should be remembered that there are other methods of optimization.

3.6.1 The multi-objective genetic algorithm

The general class of genetic algorithms (GAs) to which the MGA belongs is a class of stochastic procedures based on the concepts of natural selection in genetics. The general concept is that of evolution of populations of some sort. It acts on a population of some given size—these are the available solutions produced by the electromagnetic model. Figure 3.30 shows a simplified diagram of the process (the optimization algorithm is the shaded block on the right side of the generalized optimization process

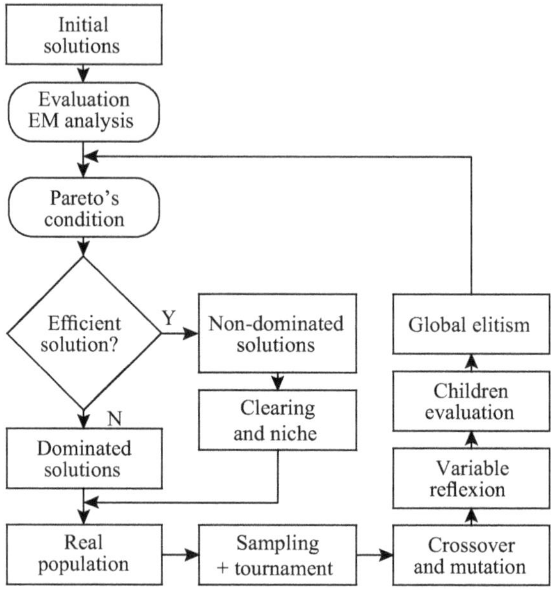

Figure 3.30 Multi-objective genetic algorithm (MGA)

in Figure 3.29). The terms associated with the MGA such as mutation and crossover, children, domination, and competition should immediately recall genetics.

In any real-world problem, be it the operation of a beehive or the optimization of an antenna, there are multiple desired goals that must be satisfied simultaneously in order to obtain an optimal solution. As these objectives are almost always conflicting, no single solution may exist that is best regarding all considered criteria. Multi-objective optimization (also called multi-criteria, multiperformance, or vector optimization) seeks to optimize the components of a vector-valued cost function. Unlike single objective optimization, the solution to this problem is not a single point but a family of efficient points from which a selection must be made. Each point in this set is optimal in the sense that no improvement can be made in a cost vector component that does not lead to degradation in at least one of the remaining components. Each element in the efficient set constitutes a non-dominated (non-inferior or non-superior) solution to the multi-objective problem.

The main action of the multi-objective optimization is to determine the efficiency front. This front or set of solutions is called the Pareto front. With this set of solutions, it is possible to understand the interdependence between objectives, leading to efficient choices for the final solution decision.

The analysis of the behavior of the Pareto-front is an important tool in understanding the trade-off between the different objectives. Compared with deterministic optimization methods, which lead to unique solution, MGAs offer the possibility to the designer to make the final choice among the set of solutions by considering additional constraints not included in the initial steps [95,97].

The MGA described here is based on three current populations. The algorithm starts with a set of randomly generated solutions. These solutions are evaluated and the Pareto-optimal condition is tested, leading to two groups of solutions: one formed by efficient solutions, called non-dominated population and another by non-efficient solutions, called dominated population.

An index indicating how many times each solution is dominated by others is created. After the Pareto set check, a clearing process is initiated, the purpose of which is to obtain a sparse and well-established Pareto front. If similarities among individuals are detected in parameters or/and objective spaces [98], one or more of them are "punished" by moving the penalized individual to the class of dominated (inefficient) population. This approach simplifies the attainment of a well-established Pareto set. Crossover and mutation operators based on coding representations are then applied to create the children of a generation. To ensure that all design variables remain inside the feasible bounds and are not affected by the evolutionary process, a procedure of "adjustment by saturation" is applied whereby design variables outside their prescribed limits are automatically adjusted to the limit values. The children are evaluated and associated with the non-dominated solution to form the new initial population. This elitism process guarantees the preservation of efficient parent solutions. The process is iterated until an ending criterion is met (typically a fixed number of generations).

3.6.2 Examples of optimization

Different GPR applications require different antennas. In addition, some applications may need improved antennas to cope with target characteristics, that is, the design of the antenna may be influenced by what the radar is looking for. Sometimes too, improvement in one antenna parameter may lead to deterioration in another. A simple example is the antenna gain, which is associated with the antenna aperture and its size. It may not be possible to obtain maximum gain while, at the same time designing the smallest antenna. A trade-off is almost always necessary.

To be able to present a meaningful review of GPR antenna design and optimization, and because there is no universal set of objectives appropriate for all GPR configurations, the antenna optimization process in this section is demonstrated using the MGA process with three objectives, all subject to two constraints. The purpose is to come up with a wideband antenna with fixed electric characteristics such as impedance. The objectives selected are as follows:

1. to maximize the directivity (D) and the gain (G), keeping the lowest HPBW possible (in relation to the normal plane of the antenna);
2. to search for a constant antenna impedance (it must be a real value, preferably close to a convenient matching impedance such as 50 or 150 Ω); and
3. to minimize the total antenna volume.

In effect, the list shows that this is an objective optimization (directivity, gain, BW, impedance, and volume). To simplify the process, and most importantly, to be able to show the trade-offs between the objectives in simple graphical plots of the Pareto set, the problem is reduced to three objectives. To do so, the directivity, gain, and bandwidth are first normalized with respect to a convenient value (such as the center value for BW or expected gain for the gain function). The three normalized values are added into a single value and the optimization proceeds with three objectives: volume, impedance, and $(D + G + \mathrm{BW})$.

The optimization is performed under two constraints:

1. VSWR (S_{11}) below -10 dB and
2. group delay (S_{12}) below -10 dB.

Other features may be included as objectives, and additional constraints may be imposed in the optimization for specific GPR applications. To find the best behavior of an antenna within a desired frequency range, and using the constraints (merit figures) (1) and (2), the algorithm considers the sum (other metrics are possible) of the evaluations in appropriate frequency steps. The number of steps depends on the antenna geometry complexity and frequency. In the following, we use the MGA to optimize some of the more common GPR antennas. The process is general and may be applied to other antennas, although the objectives and constraints may be different. One should also be aware of the fact that constraints and objectives may not be contradictory.

The following subsections describe the optimization of five different low-profile wideband antennas, as examples to the optimization process. The antennas are the rounded planar bowtie, Archimedean spiral, equiangular spiral, Vivaldi, and U-slot patch antennas [5].

To do so, the MGA was used with a population of 50 individuals (i.e., 50 sample solutions) for each of the antennas, a mutation of rate of 0.05% and crossover rate of 90%. The parameters of the antennas are adjusted between minimum and maximum limits. The number of parameters (or variables) for each antenna may be different according to each structure. Clearly, antennas with more variables may require longer processing time for the optimization and in some cases, larger populations. The initial population (solutions) is generated randomly within the limits of the variables by the electromagnetic model using either the FDTD, the transmission line method, the MoM, or the FEM (see discussion of these methods in Chapter 5). This initial population is then evaluated for each antenna based on the objectives. The solution space is then used to create the Pareto set (also called a Pareto wall). Since the objectives are often conflicting, one cannot assume that there is a "better solution" but rather that there is a group of non-dominated solutions. If, for example, one selects a better gain the volume will necessarily increase. The user must choose the appropriate solution from the Pareto set according to the needs and the application.

The discussion that follows is general in the sense that the optimization process can be applied to any antenna but, at the same time, it is tied to the specific antenna because limits on parameters and, indeed, the parameters themselves are antenna dependent. Following the discussion on these antennas, two additional examples of the classical bowtie antenna are optimized for the specific application of detection of rebar in concrete. The purpose is to show that optimization can be very specific and application oriented and, of course, that the choice of optimization objectives lies with the designer.

3.6.2.1 Rounded bowtie antenna

The rounded bowtie antenna is a planar bowtie with rounded edges and center feed. The geometry is shown in Figure 3.31. Optimization parameters are the total length l of the two wings and the flare angle ϕ, under constraints (1) and (2) (S_{11} and S_{21}) for an impedance match reference of 50 Ω. The limits on length are between 0.01 and 0.3 m, and on the flare angle between 10° and 85°. The limits on frequency are between 825 and 975 MHz. Figure 3.32 shows a 3D plot of the Pareto set for a frequency band between 825 and 975 MHz. Optimization objectives are the normalized volume, antenna impedance, and the sum of normalized values of gain, directivity, and bandwidth. The set of possible solutions is first generated using a convenient numerical method such as the FDTD method or the MoM (see Chapter 5). In this case, the MoM was used.

2D projections of the Pareto set (Figures 3.33–3.35) show the usefulness of the MGA optimization algorithm. It should be noted, however, that visualization of Pareto sets, which is relatively simple with a small number of objectives, can become complex and computation slow as the number of objectives increases.

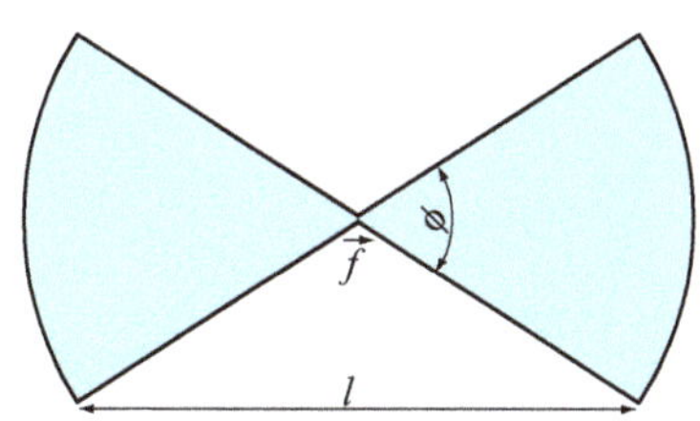

Figure 3.31 Rounded bowtie antenna indicating the length and flare angle used in the optimization ($\vec{f}$ corresponds to the location of the feed)

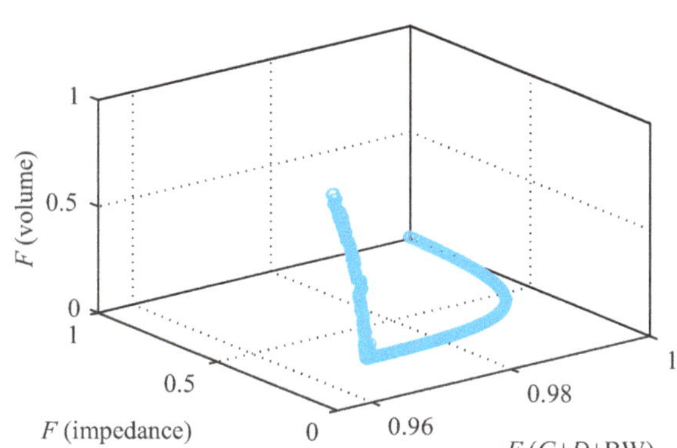

Figure 3.32 Rounded bowtie antenna—3D Pareto set

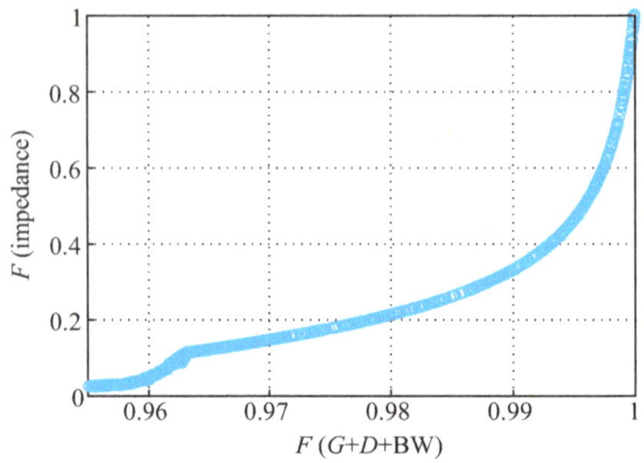

Figure 3.33 Rounded bowtie antenna—2D Pareto set: impedance versus (G+D+BW)

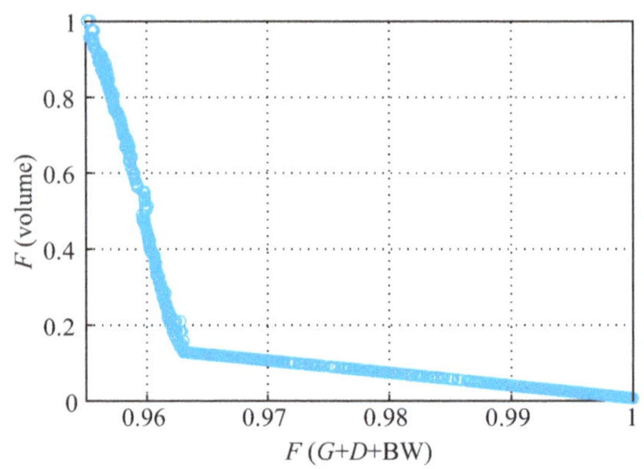

Figure 3.34 Rounded bowtie antenna—2D Pareto set: volume versus (G+D+BW)

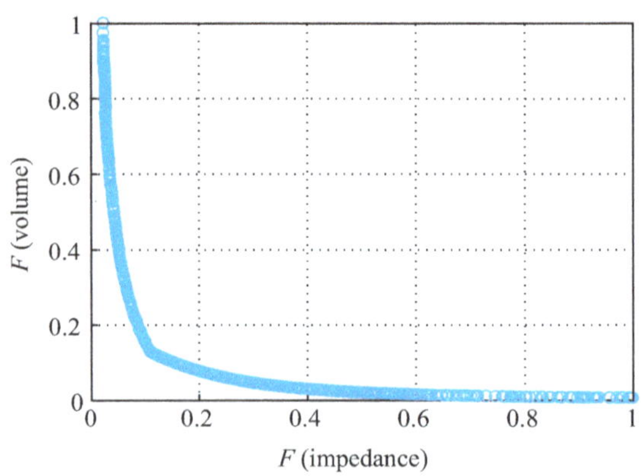

Figure 3.35 Rounded bowtie antenna—2D Pareto set: volume versus impedance

3.6.2.2 Archimedean spiral antenna

The Archimedean spiral antenna is shown in Figure 3.36. It is a planar antenna with center feed and two arms. A realizable spiral has finite limits on the feed region (e.g., distance between the two spirals, shape, and dimensions of the feed section) and on the outermost point of any arm of the spiral. The spiral antenna exhibits a broadband behavior, where the outer radius imposes the low frequency limit and the inner radius the high frequency one. The arms' radius grows linearly as a function of the winding angle. In this case, optimization parameters are the number of spiral turns N, the inner radius r_i, and the outer radius r_o. The impedance match chosen is 150 Ω. The limits for optimization are as follows:

1. number of turns: between 1.5 and 2.5;
2. inner radius r_i: between 4 and 5.5 mm;
3. outer radius r_o: between 37.5 and 50 mm; and
4. frequency: between 825 and 975 MHz.

Figure 3.37 shows the 3D Pareto set obtained in the optimization. In comparison with the Pareto set for the bowtie antenna (Figure 3.37), it should be noted that the set in Figure 3.37 is not smooth, with outlier points. These indicate the way MGA

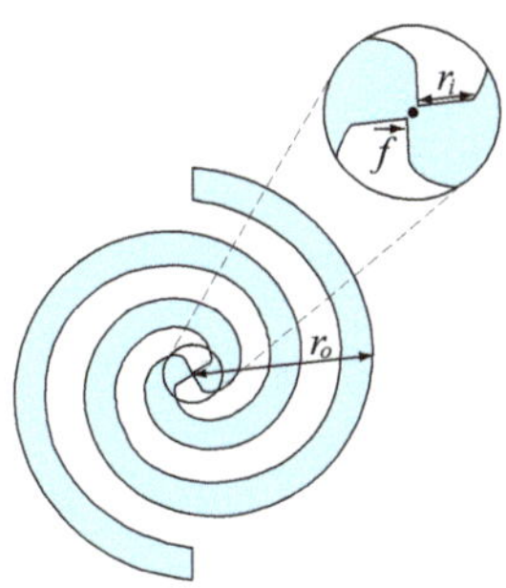

Figure 3.36 Archimedean spiral antenna ($\vec{f}$ corresponds to the location of the feed)

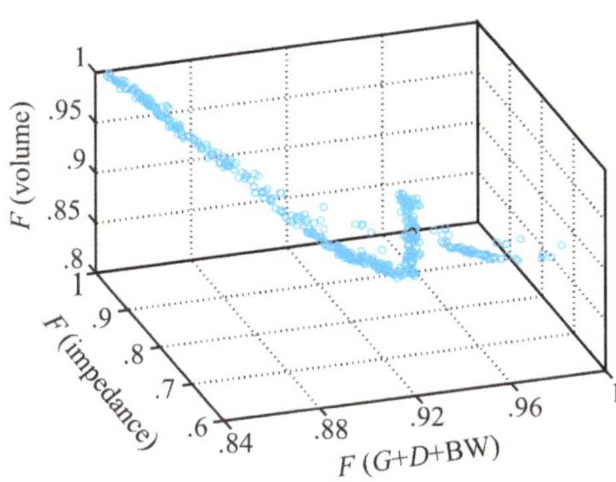

Figure 3.37 Archimedean spiral antenna: 3D Pareto set

performs in that it can identify outliers and correct for these to eventually converge to acceptable solutions.

3.6.2.3 Equiangular spiral antenna

The equiangular spiral antennas shown in Figure 3.38 show noticeable differences in comparison with the Archimedean spiral, with the main difference in that this antenna is viewed as FI. An antenna based on a log-spiral curve can be specified entirely by angles except for a necessary arm length, forming thus an FI structure. Optimization parameters in this case are the inner radius r_i, the outer radius r_o, and a growth rate imposed on the spirals that defines their total length. The impedance match chosen is again 150 Ω, with the same constraints on S_{11} and S_{21}. The limits for optimization are as follows:

1. Growth rate: between 0.35 and 0.5;
2. inner radius r_i: between 4 and 5.5 mm;
3. outer radius r_o: between 37.5 and 50 mm; and
4. frequency: between 825 and 975 MHz.

Figure 3.39 shows the 3D Pareto set. Comparing the 3D Pareto sets for the Archimedean and equiangular spiral antennas (Figures 3.37 and 3.39), it can be seen

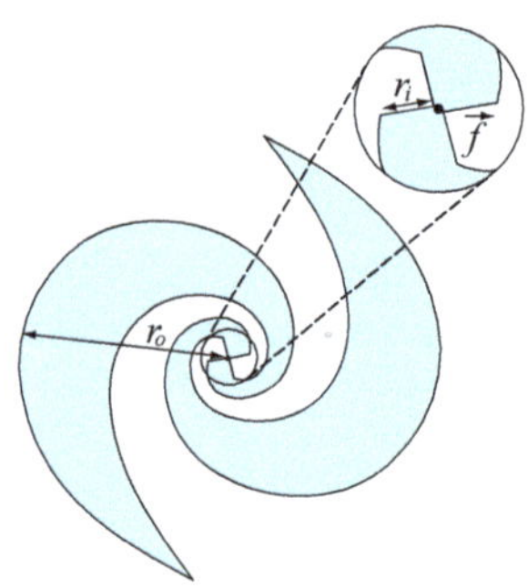

Figure 3.38 Equiangular spiral antenna ($\vec{f}$ corresponds to the location of the feed)

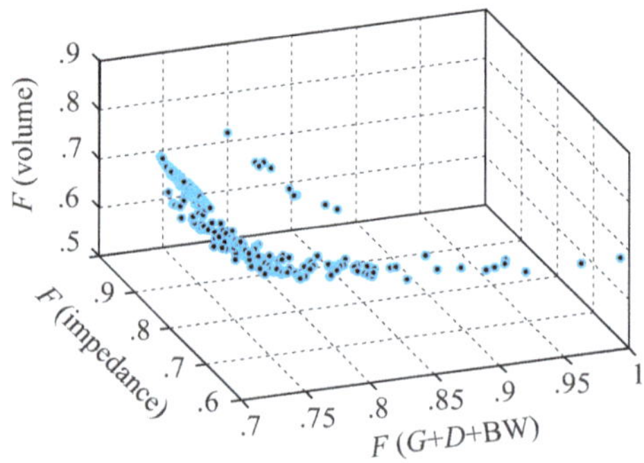

Figure 3.39 Equiangular spiral antenna: 3D Pareto set

that the Archimedean spiral antenna has better optimization values than the equiangular spiral antenna. The main reason is that the arms are closer for the Archimedean geometry, resulting in smaller volumes and better radiation characteristics. Another reason is that for a band-limited operation, which in this case is rather narrow, an FI antenna's most attractive feature has limited appeal and limited effect on the optimization.

3.6.2.4 Vivaldi antenna

The Vivaldi notch antenna in Figure 3.40 on a ground plane is also known as a tapered slot antenna. Power is radiated from the aperture and reaches an exponentially tapered pattern by using a symmetrical slot line. The optimization parameters are the tapered length, aperture width, opening angle, cavity total tapered spacing, ground plane length, and ground plane width with the following limits:

1. opening rate: between 20° and 30°;
2. taper length l_S: between 24.2 and 24.5 cm;
3. aperture width u: between 10 and 11 cm;
4. ground plane length l: between 29.5 and 30.5 cm;
5. ground plane width w: between 12 and 13 cm; and
6. frequency: between 825 and 975 MHz.

Other dimensions are fixed: slotline width (W_S) is 0.5 mm, cavity diameter (d) is 2.4 cm, and cavity to taper spacing (S) is 2.3 cm. Optimization is done for a reference impedance of 50 Ω. Figure 3.41 shows the 3D Pareto sets for the Vivaldi antenna. The antenna can operate at high frequencies, with very stable radiation parameters in the whole bandwidth. Note also that the limits for the various optimization parameters were set to rather narrow ranges. This was done here for the sole purpose of obtaining solutions in reasonably short time. Wider limit ranges may reach better solutions at the expense of additional processing time. This trade-off should be kept in mind in any optimization.

3.6.2.5 U-slot patch antenna

The U-slot patch antenna consists of a rectangular patch radiator within a U-shaped slot (Figure 3.42). The patch itself is on an air substrate which is sufficiently thick to achieve wide bandwidths. The slot structure provides additional capacitances in

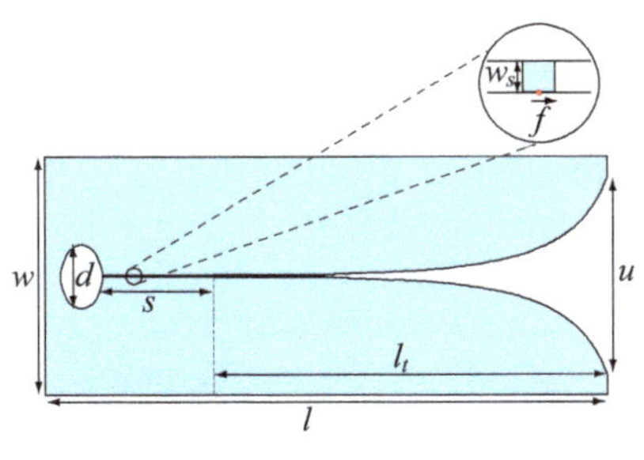

Figure 3.40 Vivaldi antenna ($\vec{f}$ corresponds to the location of the feed)

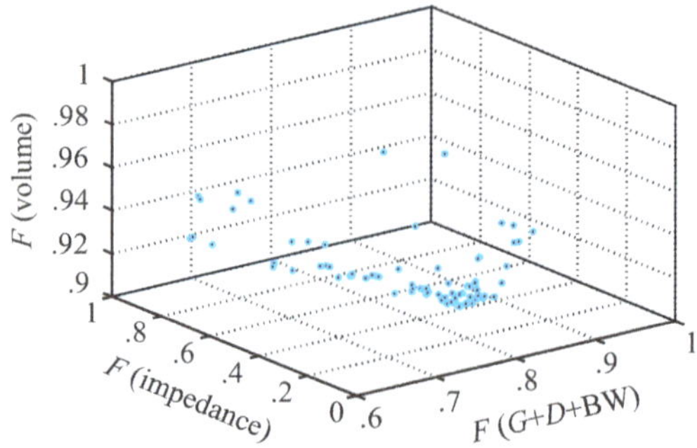

Figure 3.41 Vivaldi antenna: 3D Pareto set

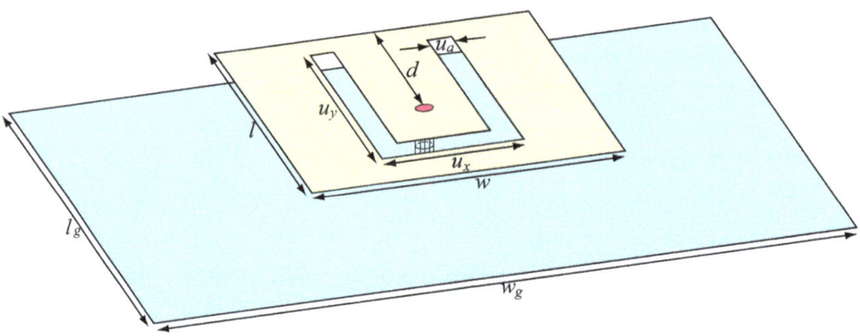

Figure 3.42 U-slot patch antenna

the structure, which, combined with the inductance of the long probe feed creates a double resonance in the band. The optimization parameters in this case are the size of the rectangular patch (length and width), the size of the U-slot (length, width, opening, and distance from the rectangular patch edge), the geometry of the feed (distance from the rectangular patch edge and height from the ground plane), and size of the ground plane (length and width). The impedance reference is 50 Ω.

1. length of patch l: between 25.2 and 26.5 mm;
2. width of patch w: between 35.25 and 35.75 mm;
3. aperture length u_y: between 19 and 20 mm;
4. aperture width u_x: between 11 and 13 mm;
5. aperture opening u_a: between 2 and 2.2 mm;
6. length of ground plane l_g: between 50 and 55 mm;
7. width of ground plane w_g: between 70 and 72 mm;
8. distance of feed from edge d: between 13 and 14 mm;
9. distance between ground plane and patch d: between 5.5 and 6.5 mm; and
10. frequency: between 825 and 975 MHz.

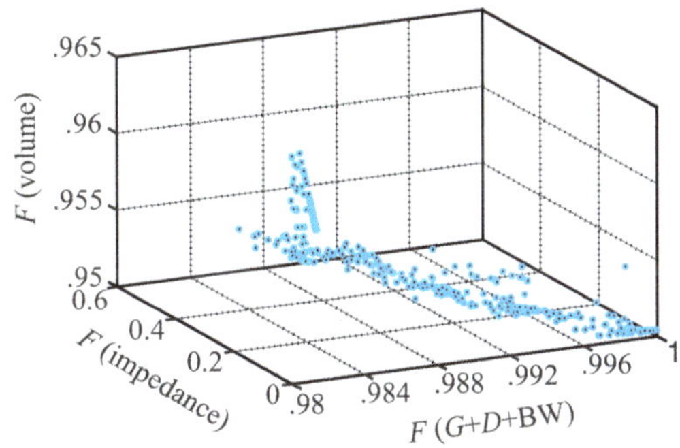

Figure 3.43 U-slot patch antenna: 3D Pareto set

As with the Vivaldi antenna, some parameters are fixed. For example, the horizontal distance of the patch from the edge of the ground plane is 4.8 mm. Note also that the range between minimum and maximum values is rather small. There is no rule by which these values may be selected except for the fact that initial, starting values must be supplied. However, when the number of parameters is large (as in this case), the larger the span of the parameters, the larger the number of solutions that must be generated for the MGA optimization to be able to select meaningful solutions. This increases both the time and cost of electromagnetic simulations in the initial solution generation step and that of the optimization process because of the larger number of possible solutions. On the other hand, if the spans are large and the number of solutions too small to adequately cover the spans, the optimization may not be adequate or may fail entirely. Figure 3.43 shows the 3D Pareto set for the U-slot patch antenna.

In comparison with the Vivaldi antenna, the U-slot patch shows slightly better averaged values. However, the U-slot antenna is more complex and the issues of fabrication of the optimized antenna must be taken into account.

3.6.3 Optimization for specific applications

The optimization of the antennas in the previous sections was general in that the antennas were radiating into a space without the antenna being affected by any particular body or material properties. Also, the choice of optimization parameters were selected to be generic in the sense that they would apply to many types of antennas. It was also mentioned that in many cases, the optimization can and should take into account the actual geometry and type of assessment that must be done because by doing so one can improve performance for a particular application or class of applications. A good, general example is assessment of concrete. Although this is a very specific application, it is so widespread that it justifies efforts to improve performance taking into account, for example, the properties of concrete. Another example is mine detection where the critical importance of the application justifies any effort to improve performance, taking into account the particulars of mine detection. The following section discusses the optimization of two bowtie antennas for the assessment of features in concrete. One is a classic planar bowtie, whereas the second is a V-shaped bowtie in which the two wings of the bowtie form an angle with respect to the normal to the antenna.

3.6.3.1 Planar bowtie antenna optimization

The design of an effective planar bowtie antenna requires balancing the antenna length, the flare angle, and the radiation pattern produced (or, alternatively, the gain of the antenna, as was done, e.g., in Section 3.6.2.1). Most antenna characteristics that are relevant to GPR applications such as wave polarization, radiation field pattern, and bandwidth are commonly defined in the far-field region of the antenna. However, notwithstanding the complexity of the electromagnetic radiation in the near-field region, most civil engineering applications using surface contact antennas are concerned with radar measurements in the near field [99]. Thus, consideration should be given to the effects on the antenna in the near field as well, where the proximity of surfaces and, in some cases, of targets, can affect antenna performance.

The discussion starts with a planar dipole in free space followed by optimization in the presence of concrete with and without rebar. The goal in the optimized design of this antenna is to reduce the metal area (and consequently to reduce the weight of the antenna) and to improve the gain in the plane perpendicular to the antenna. The MGA is coded to find multiple non-dominated solutions (the Pareto-front) using a fixed frequency of 1 GHz. The antenna parameters to adjust are as follows:

$$P^g = \begin{bmatrix} L^{g,1} & \alpha^{g,1} & E^{g,1} \\ \vdots & \vdots & \vdots \\ L^{g,np} & \alpha^{g,np} & E^{g,np} \end{bmatrix} \tag{3.43}$$

where each line represents a feasible solution, g is the current generation, and np is the population size (number of simulations). The variables to be optimized are the antenna length, the flare angle, and the percentage of antenna elements that can be removed (replaced with air). The latter are adjusted to minimize the metal area of the antenna. This becomes the first objective function. The second objective function is to maximize the gain in the direction perpendicular to the plane of the antenna.

In order to find the antenna configuration with the highest directivity and the smallest metal area, the MGA is required to accomplish two conflicting objectives with the following limits: length L [0.1λ–1λ] (with frequency equal to 1 GHz), flare angle [30°–120°], and with appropriate void spaces in the antenna structure to minimize the metal area (weight).

The antenna is initially created with 256 triangular elements and then a percentage of the elements between 0 and 20 is changed from metal to air according to the objectives. The feed region is obviously protected to avoid numerical errors and nonphysical solutions.

Figure 3.44 shows the Pareto front, clearly indicating the trade-off between metal surface and gain. The figure also shows the point in the Pareto front selected as the optimized design. The antenna obtained at that point in the Pareto front is shown in Figure 3.45.

The optimized antenna suggested by the algorithm with a maximum gain has a flare angle α = 79° and length L = 26 cm with 11% of the elements erased. The radiation pattern in Figure 3.46 shows the gain obtained in the plane normal to the antenna for the design in Figure 3.45. The gain is 6.37 dB compared to 3.40 dB of the common structure. The HPBW improves from 57.6° to 43.2°. Other solutions can be found to satisfy other design needs.

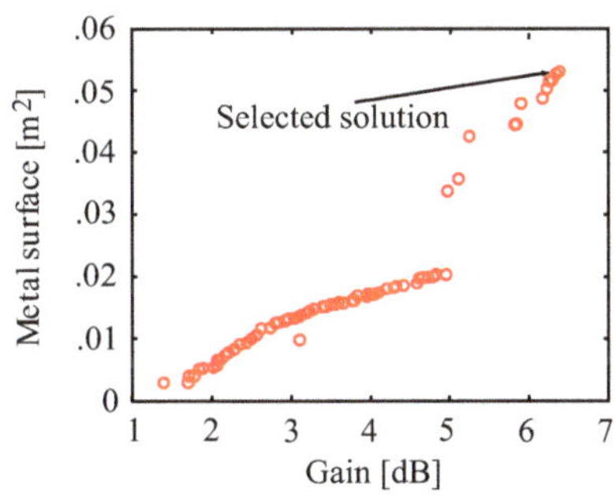

Figure 3.44 Pareto front for the planar bowtie antenna showing the selected optimized solution

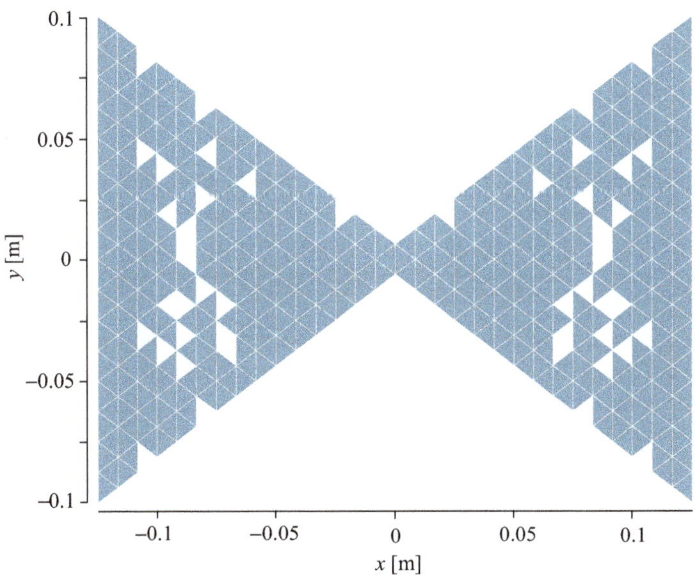

Figure 3.45 Optimized planar bowtie antenna

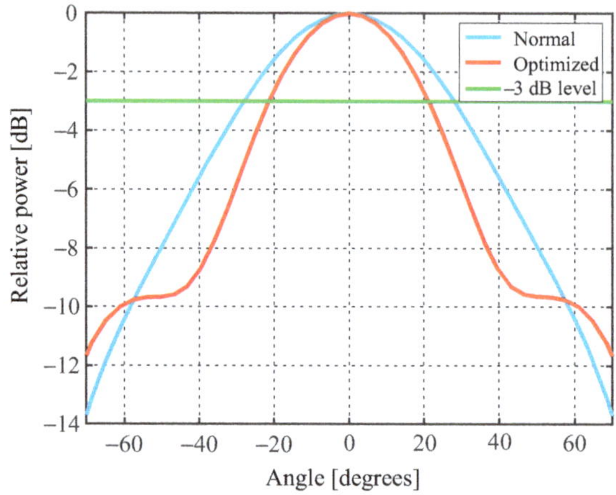

Figure 3.46 H-plane field pattern of the planar bowtie antenna

Table 3.4 Data for concrete

Porosity of concrete	0.15
Degree of saturation	0.7
Salt content	52 ppt
Temperature	20°C

Convergence has been attained in about 50 generations with a population of 50 individuals in a number of repeated GA executions. The crossover and mutation probabilities were set to 0.9 and 0.05, respectively.

3.6.3.2 V-shaped bowtie antenna optimization

To take into account the coupling effects of the antenna when on (or in close proximity with) a dielectric interface, a more realistic model of the concrete structure was implemented and solutions for the optimization were obtained using the FEM—see Chapter 5. The antenna is modeled and optimized in the presence of concrete with a rebar embedded in it.

For the electrical properties of concrete, a discrete model was used to compute the complex permittivity for each frequency component [100]. The electrical properties obtained for concrete are shown in Table 3.4. In addition, a conducting shield was added to the antenna to improve directivity. At 1 GHz, the concrete slab was simulated with $\varepsilon_r = 8.37$ and $\sigma = 0.23$ S/m. A steel rebar, buried 15 cm deep and parallel or perpendicular to the plane of the antenna, is the target to be detected.

Starting with the flat bowtie, the angle between the bowtie wings was added as a new variable producing a V-shaped bowtie antenna. The angle was allowed to vary from 180° (flat bowtie) to 45°. The Pareto-front for the V-shaped bowtie antenna is shown in Figure 3.47 with the last point in the Pareto front selected as the optimized

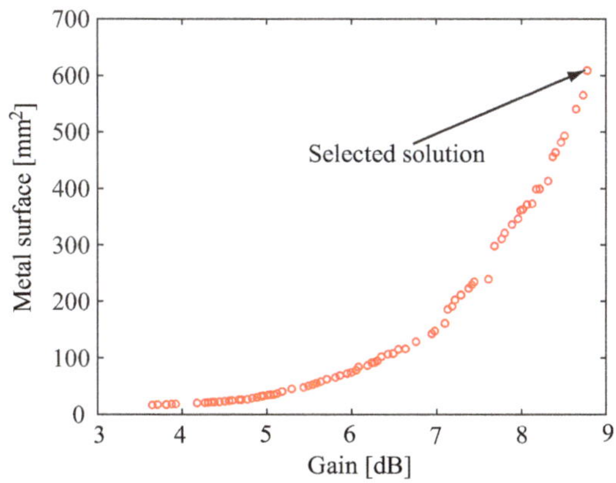

Figure 3.47 Pareto front for the V-shaped bowtie antenna with the optimized solution indicated

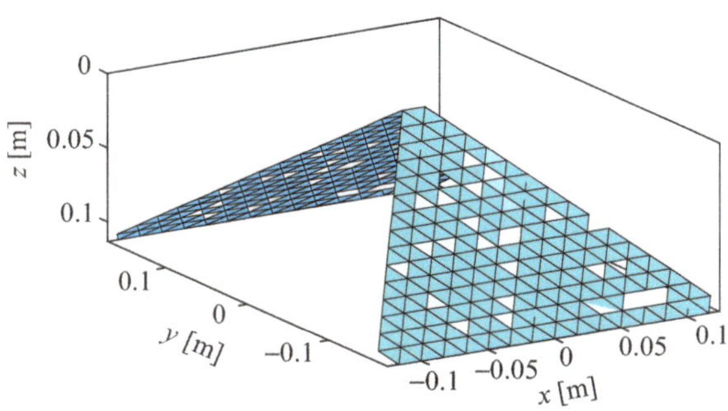

Figure 3.48 V-shaped antenna as obtained from the Pareto front in Figure 3.47

solution (indicated by the arrow). The overall requirement here was to obtain a solution with a maximum gain minimizing the importance of the weight. The resulting antenna is shown in Figure 3.48. Its radiation pattern is shown in Figure 3.49. Sidelobes are minimal and radiation is perpendicular to the antenna.

To demonstrate additional possibilities in the optimization, a new constraint was imposed on the design: the return loss of the antenna. Antennas with return loss greater than −10 dB for a transmission line feed with a characteristic impedance of 200 Ω were penalized in the optimization process. The gain obtained was of 8.77 dB with a return loss of −13.5 dB at 1 GHz. The angle between the bowtie wings for this design was 97.88°. In this case, the same antenna without the removed metal would not fulfill the impedance criteria.

Figure 3.50 shows the modifications in the antenna's input impedance for three different scenarios. In the case where the bar is perpendicular to the antenna, and consequentially, located in a region of high illumination, the input impedance is highly affected indicating its presence.

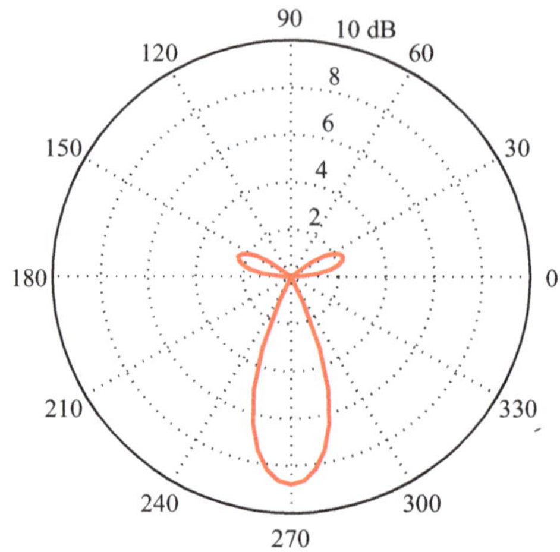

Figure 3.49 Radiation pattern for the V-shaped bowtie antenna

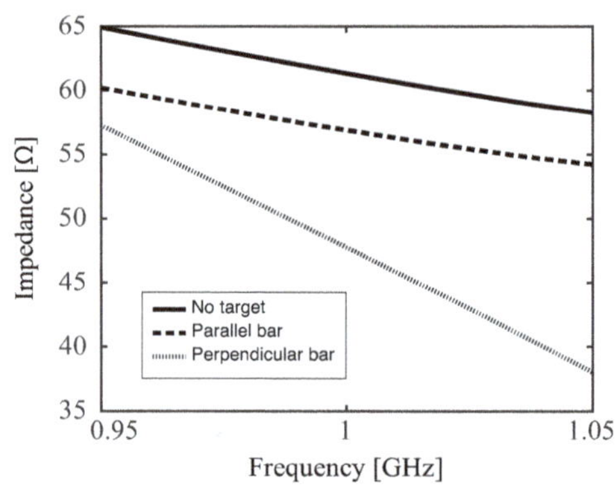

Figure 3.50 Input impedance of the V-shaped antenna in the radar assessment

Finally, the antenna impedance is shown in Figure 3.51 as a function of frequency over the range 0.5–1.5 GHz, indicating a fairly constant impedance with a radiation resistance of just under 200 Ω.

3.6.3.3 Some remarks on optimization

Although the discussion on optimization of antennas was limited in scope, some conclusions can nevertheless be advanced. For optimization purposes, the antenna can be considered as a two-port transducer system characterized by a transfer function that can be frequency dependent and may have a nonlinear phase response. The performance of the antenna is critical for both time- and frequency-domain GPR systems. The system configuration directly affects the performance of the NDT assessment and requires different characteristics and features from antennas. The examples

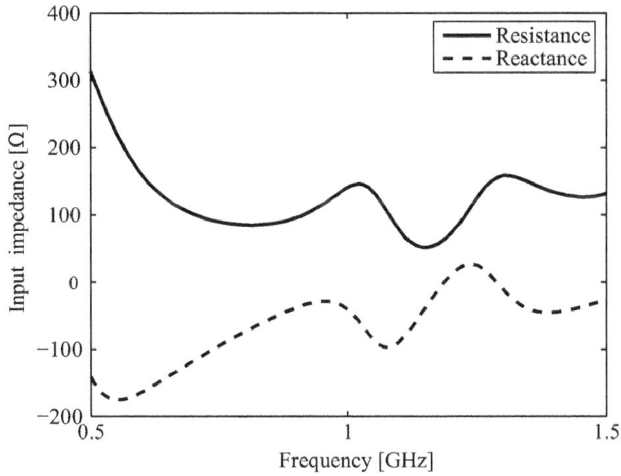

Figure 3.51 V-shaped antenna impedance as a function of frequency

shown here outline some fundamentals of the GPR system theory in order to translate the GPR system configuration into antenna requirements. Although different GPR configurations impose different restrictions on antenna type and design, the procedure outlined here can help one to achieve high-performance antennas, is consistent, and can be described as a generalized optimization problem. The analyzed antennas illustrate the computational burden in using multi-objective optimization algorithms to enhance their performance. When possible, stochastic-based optimization tools should be used since they are better suited than deterministic methods. The multi-objective algorithm was used to show that stochastic methods are able to provide meaningful sets of efficient solutions in the presence of targets or in space. A proper acceptable design for any application requires a set of criteria to allow the designer to choose an optimal solution from the Pareto front.

References

[1] IEEE. IEEE Standard for Definitions of Terms for Antennas. In IEEE Std 145-2013 (Revision of IEEE Std 145-1993), 2014:1–50. doi: 10.1109/IEEESTD.2014.6758443.

[2] Licul S and Davis WA. Unified frequency and time-domain antenna modeling and characterization. IEEE Transactions on Antennas and Propagation, 53 (9); 2005:2882–2888.

[3] Shlivinski A, Heyman E, and Kastner R. Antenna characterization in the time domain. IEEE Transactions on Antennas and Propagation, 45 (7);1997: 1140–1149.

[4] Farr E and Baum C. Time Domain Characterization of Antennas with TEM Feeds. Air Force Research Laboratory, USA; 1998. note 426.

[5] Balanis C. Antenna Theory—Analysis and Design. John Wiley & Sons, Hoboken, NJ; 2016.
[6] Smith G. Introduction to Classical Electromagnetic Radiation. Cambridge University Press, Cambridge; 1997.
[7] Schantz HG. Dispersion and UWB antennas. In: 2004 International Workshop on Ultra Wideband Systems Joint with Conference on Ultra Wideband Systems and Technologies. Joint UWBST IWUWBS 2004 (IEEE Cat. No.04EX812); 2004:161–165.
[8] Neal A. Ground-penetrating radar and its use in sedimentology: Principles, problems and progress. Earth-Science Reviews, 66 (3);2004:261–330.
[9] Born M and Wolf E. Principles of Optics, 7th ed. Cambridge University Press, Cambridge; 2005.
[10] Wait JR. Electromagnetic surface waves. Advances in Radio Research, 1;1964:157–217.
[11] Banos A. Dipole Radiation in the Presence of a Conducting Half-Space. Pergamon, New York, NY; 1966.
[12] Annan AP. Radio interfering depth sounding. Part I: Theoretical discussion. Geophysics, 38;1973:557–580.
[13] Rutledge DB and Muha MS. Imaging antenna arrays. IEEE Transactions on Antennas and Propagation, 30;1982:535–540.
[14] Smith GS. Directive properties of antenna for transmission into a material half-space. IEEE Transactions on Antennas and Propagation, 32;1984: 232–246.
[15] Balanis CA. Modern Antenna Handbook. John Wiley & Sons, Hoboken, NJ; 2008.
[16] Cai Z, Lin W, Qi X, *et al.* Design of low-cost ground penetrating radar receiving circuit based on equivalent sampling. In: 2019 IEEE International Symposium on Circuits and Systems (ISCAS); 2019:1–4.
[17] Reisenzahn A, Buchegger T, Scherrer D, *et al.* A ground penetrating UWB radar system. In: 2006 3rd International Conference on Ultrawideband and Ultrashort Impulse Signals; 2006. p. 116–118.
[18] Travers D. The effect of mutual impedance on the spacing error of an eight-element Adcock. IRE Transactions on Antennas and Propagation, 5 (1); 1957:36–39.
[19] Lytle RJ and Laine EF. Design of a miniature directional antenna for geophysical probing from boreholes. IEEE Transactions on Geoscience Electronics, 16 (4); 1978:304–307.
[20] Shlager KL, Smith GS, and Maloney JG. A resistively loaded bowtie antenna for pulse radiation: FDTD analysis and optimization. In: Proceedings of IEEE Antennas and Propagation Society International Symposium. vol. 2; 1993:830–833.
[21] Wu T and King R. The cylindrical antenna with nonreflecting resistive loading. IEEE Transactions on Antennas and Propagation, 13 (3);1965:369–373.

[22] Liu H, Xing B, Zhu J, *et al.* Quantitative stability analysis of ground penetrating radar systems. IEEE Geoscience and Remote Sensing Letters, 15 (4); 2018:522–526.

[23] Minet J, Lambot S, Slob EC, *et al.* Soil surface water content estimation by full-waveform GPR signal inversion in the presence of thin layers. IEEE Transactions on Geoscience and Remote Sensing, 48 (3);2010:1138–1150.

[24] Rumsey V. Frequency independent antennas. In: 1958 IRE International Convention Record. vol. 5; 1957:114–118.

[25] Miller EK and Landt JA. Direct time-domain techniques for transient radiation and scattering from wires. Proceedings of the IEEE, 68 (11);1980: 1396–1423.

[26] Schantz H. The Art and Science of Ultrawideband Antennas. Second edition ed. Artech House Antennas and Propagation Library. Artech House, Boston; 2015.

[27] Ali EJ, Abdullah N, Ismail M, *et al.* Ultra-wideband antenna design for GPR applications: A review. International Journal of Advanced Computer Science and Applications, 8;2017:392–400.

[28] Pieraccini M, Rojhani N, and Miccinesi L. Comparison between horn and bow-tie antennas for ground penetrating radar. In: 2017 9th International Workshop on Advanced Ground Penetrating Radar (IWAGPR); 2017. p. 1–5.

[29] Pajewski L, Tosti F and Kusayanagi W. Antennas for GPR systems. In: Benedetto A, and Pajewski L (eds.), Civil Engineering Applications of Ground Penetrating Radar. Springer Transactions in Civil and Environmental Engineering. Springer, Cham; 2015. https://doi.org/10.1007/978-3-319-04813-0_2.

[30] Jonard F, Weihermller L, Schwank M, *et al.* Estimation of hydraulic properties of a sandy soil using ground-based active and passive microwave remote sensing. IEEE Transactions on Geoscience and Remote Sensing, 53 (6); 2015:3095–3109.

[31] Li X, Ji Y, Lu W, *et al.* Analysis of GPR antenna system mounted on a vehicle. IEEE Antennas and Wireless Propagation Letters, 12;2013:575–578.

[32] Giannakis I, Giannopoulos A, and Warren C. A realistic FDTD numerical modeling framework of ground penetrating radar for landmine detection. IEEE Journal of Selected Topics in Applied Earth Observations and Remote Sensing, 9 (1);2016:37–51.

[33] Almeida ER, Porsani JL, Catapano I, *et al.* Microwave tomography-enhanced GPR in forensic surveys: The case study of a tropical environment. IEEE Journal of Selected Topics in Applied Earth Observations and Remote Sensing, 9 (1);2016:115–124.

[34] Höfinghoff J and Overmeyer L. Resistive loaded antenna for ground penetrating radar inside a bottom hole assembly. IEEE Transactions on Antennas and Propagation, 61 (12);2013:6201–6205.

[35] Liu Q, Wang Y, and Fathy AE. A compact integrated 100 GS/s sampling module for UWB see through wall radar with fast refresh rate for dynamic real time imaging. In: 2012 IEEE Radio and Wireless Symposium; 2012:59–62.

[36] De Pue J, Van Meirvenne M, and Cornelis WM. Accounting for surface refraction in velocity semblance analysis with air-coupled GPR. IEEE Journal of Selected Topics in Applied Earth Observations and Remote Sensing, 9 (1); 2016:60–73.

[37] Zeng Z, Li J, Huang L, *et al.* Improving target detection accuracy based on multipolarization MIMO GPR. IEEE Transactions on Geoscience and Remote Sensing, 53 (1);2015:15–24.

[38] De Coster A, Tran AP, and Lambot S. Fundamental analyses on layered media reconstruction using GPR and full-wave inversion in near-field conditions. IEEE Transactions on Geoscience and Remote Sensing, 54 (9);2016:5143–5158.

[39] Guan B, Ihamouten A, Dérobert X, *et al.* Near-field full-waveform inversion of ground-penetrating radar data to monitor the water front in limestone. IEEE Journal of Selected Topics in Applied Earth Observations and Remote Sensing, 10 (10);2017:4328–4336.

[40] Salucci M, Oliveri G, Rocca P, *et al.* A system-by-design approach for efficient multiband patch antennas design. In: 2017 International Applied Computational Electromagnetics Society Symposium – Italy (ACES); 2017. p. 1–2.

[41] Maiti S, Patra SK, and Bhattacharya A. A modified plane wave model for fast and accurate characterization of layered media. IEEE Transactions on Microwave Theory and Techniques, 65 (9);2017:3492–3502.

[42] Comite D, Galli A, Lauro SE, *et al.* Analysis of GPR early-time signal features for the evaluation of soil permittivity through numerical and experimental surveys. IEEE Journal of Selected Topics in Applied Earth Observations and Remote Sensing, 9 (1);2016:178–187.

[43] Kim K and Scott WR Jr. In: Harmon RS, Broach JT, and Holloway JH Jr, editors. A Resistive Linear Antenna for Ground-Penetrating Radars. Orlando, FL: Defense and Security; 2004. p. 359.

[44] Lestari AA, Yarovoy AG, and Ligthart LP. RC-loaded bow-tie antenna for improved pulse radiation. IEEE Transactions on Antennas and Propagation, 52 (10);2004:2555–2563.

[45] Kim K and Scott WR. Design and realization of a discretely loaded resistive Vee dipole for ground-penetrating radars. Radio Science, 39 (4);2004:1–9.

[46] Lestari AA, Yulian D, Liarto, *et al.* Theoretical and experimental analysis of a rolled-dipole antenna for low-resolution GPR. In: 2007 IEEE International Conference on Ultra-Wideband; 2007:294–298.

[47] Lestari AA. Antennas for Improved Ground Penetrating Radar: Modeling Tools, Analysis and Design. Technische Universiteit Delft; 2003.

[48] Lestari AA, Yarovoy AG, and Ligthart LP. Numerical and experimental analysis of circular-end wire bow-tie antennas over a lossy ground. IEEE Transactions on Antennas and Propagation, 52 (1);2004:26–35.

[49] Gibson PJ. The Vivaldi aerial. In: 1979 9th European Microwave Conference; 1979:101–105.

[50] Thaysen J, Jakobsen KB, and Lenler-Eriksen HR. Wideband cavity backed spiral antenna for stepped frequency ground penetrating radar. In: 2005 IEEE Antennas and Propagation Society International Symposium. vol. 1B; 2005:418–421.

[51] Dyson J. The equiangular spiral antenna. IRE Transactions on Antennas and Propagation, 7 (2);1959:181–187.

[52] Schrank H. Antenna designer's notebook. IEEE Antennas and Propagation Society Newsletter, 31 (4);1989:25–28.

[53] Shao J, Fang G, Fan J, *et al.* TEM horn antenna loaded with absorbing material for GPR applications. IEEE Antennas and Wireless Propagation Letters, 13;2014:523–527.

[54] Ahmed A, Zhang Y, Burns D, *et al.* Design of UWB antenna for air-coupled impulse ground-penetrating radar. IEEE Geoscience and Remote Sensing Letters, 13 (1);2016:92–96.

[55] Yarovoy AG, Schukin AD, Kaploun IV, *et al.* The dielectric wedge antenna. IEEE Transactions on Antennas and Propagation, 50 (10);2002:1460–1472.

[56] Lubetsky NV and Palto AA. Optimization of antenna spacing in antenna array for ground penetrating radar. In: 2005 15th International Crimean Conference Microwave Telecommunication Technology. vol. 2; 2005:941–942.

[57] Visser HJ. Array and Phased Array Antenna Basics: Visser/Array and Phased Array Antenna Basics. John Wiley & Sons, Ltd, Chichester, UK; 2005.

[58] Das U, Boer HJ, and van Ardenne A. Phased array technology for GPR antenna design for near subsurface exploration. In: Proceedings of the 2nd International Workshop on Advanced Ground Penetrating Radar; 2003:30–35.

[59] Utsi EC. Ground Penetrating Radar: Theory and Practice. Elsevier Science, Cambridge, MA; 2017.

[60] Jol HM. Ground Penetrating Radar Theory and Applications. Elsevier Science, Amsterdam; 2008.

[61] Wong PTW, Lai WWL, and Sato M. Time-frequency spectral analysis of step frequency continuous wave and impulse ground penetrating radar. In: 2016 16th International Conference on Ground Penetrating Radar (GPR); 2016: 1–6.

[62] Fioranelli F. Through-The-Wall Detection Using Ultra Wide Band Frequency Modulated Interrupted Continuous Wave Signals. Durham University; 2013.

[63] Salucci M, Poli L, Anselmi N, *et al.* Multifrequency particle swarm optimization for enhanced multiresolution GPR microwave imaging. IEEE Transactions on Geoscience and Remote Sensing, 55 (3);2017:1305–1317.

[64] Busch S, van der Kruk J, and Vereecken H. Improved characterization of fine-texture soils using on-ground GPR full-waveform inversion. IEEE Transactions on Geoscience and Remote Sensing, 52 (7);2014:3947–3958.

[65] Deng R and Liu C. FM-CW radar performance in a lossy layered medium. Journal of Applied Geophysics, 42 (1);1999:23–33.

[66] Kong FN and By TL. Performance of a GPR system which uses step frequency signals. Journal of Applied Geophysics, 33 (1);1995:15–26.

[67] Diamanti N and Annan AP. Air-launched and ground-coupled GPR data. In: 2017 11th European Conference on Antennas and Propagation (EUCAP); 2017:1694–1698.

[68] Van De Vijver E, De Pue J, Cornelis W, *et al.* Comparison of air-launched and ground-coupled configurations of SFCW GPR in time, frequency and wavelet domain. In: EGU General Assembly Conference Abstracts. vol. 17 of EGU General Assembly Conference Abstracts; 2015:10038.

[69] Electronic Code of Federal Regulations, Section 15.509 – Technical requirements for ground penetrating radars and wall imaging systems. Accessed: 2021-02-21. https://nam03.safelinks.protection.outlook.com.

[70] Electronic Code of Federal Regulations, Section 15.521 – Technical requirements applicable to all UWB devices. Accessed: 2021-02-2021. https://nam03.safelinks.protection.outlook.com.

[71] Taflove A and Hagness SC. Computational Electrodynamics: The Finite-difference Time-domain Method. Artech House Antennas and Prop. Artech House, Boston, MA; 2005.

[72] Ida N. Engineering Electromagnetics. 2nd ed. Springer Publishing Company, Incorporated, Cham, Switzerland; 2007.

[73] Sadiku MNO. Numerical Techniques in Electromagnetics with MATLAB. CRC Press; 2018.

[74] Harrington RF. Field Computation by Moment Methods. Chapman and Hall/CRC, 2014.

[75] Jin J. The Finite Element Method in Electromagnetics. Wiley-IEEE Press, 2002.

[76] Davidson DB. Computational Electromagnetics for RF and Microwave Engineering. Cambridge University Press, 2014.

[77] Warren C, Giannopoulos A and Giannakis I. gprMax: Open source software to simulate electromagnetic wave propagation for Ground Penetrating Radar. Computer Physics Communications, 209;2016:163–170. doi:10.1016/j.cpc.2016.08.020.

[78] Warren C and Giannopoulos A. Experimental and modeled performance of a ground penetrating radar antenna in lossy dielectrics. IEEE Journal of Selected Topics in Applied Earth Observations and Remote Sensing, 9 (1); 2016:29–36.

[79] MIT Electromagnetic Equation Propagation (MEEP). MEEP free finite-difference time-domain (FDTD) simulation software package to model electromagnetic systems, 2017. Accessed: 2021-21-2021. https://meep.readthedocs.io/en/latest/.

[80] ANSYS, Inc. ANSYS HFSS: High Frequency Electromagnetic Field Simulation, 2017. Accessed: 2018-04-16. http://www.ansys.com/products/electronics/ansys-hfss.

[81] Altair Hyperworks, Inc. FEKO 3D computational electromagnetic software, 2017. Accessed: 2018-04-16. http://www.altairhyperworks.com/product/FEKO.

[82] DassaultSystemes, Inc. Computer Simulation Technology – CST Microwave Studio, 2017. Accessed: 2018-04-16. https://www.cst.com/products/cstmws/solvers.

[83] GSSI, Inc. RADAN GPR Data Processing Software, 2017. Accessed: 2018-04-16. http://www.geophysical.com/software.htm.

[84] Golding P, Kapadia S, Naylor S, *et al.* Framework for minimising the impact of regional shocks on global food security using multi-objective ant colony optimisation. Environmental Modelling Software, 95;2017:303–319.

[85] Biancheri-Astier M, Ciarletti V, Reineix A, *et al.* Optimization of the antennas of the EISS radar designed to perform deep Martian subsurface Sounding. IEEE Transactions on Geoscience and Remote Sensing, 53 (8);2015:4627–4637.

[86] van Coevorden CMJ, Bretones AR, Fernández Pantoja M, *et al.* Thin-wire antenna design for GPR applications using a multi-objective GA. Near Surface Geophysics, 5 (1);2007:23–28.

[87] Li M, Birken R, Sun NX, *et al.* Compact slot antenna with low dispersion for ground penetrating radar application. IEEE Antennas and Wireless Propagation Letters, 15;2016:638–641.

[88] Nayak R, Maiti S, and Patra SK. Design and simulation of compact UWB Bow-tie antenna with reduced end-fire reflections for GPR applications. In: 2016 International Conference on Wireless Communications, Signal Processing and Networking (WiSPNET); 2016:1786–1790.

[89] Liu S, Wang Q, and Gao R. A topology optimization method for design of small GPR antennas. Structural and Multidisciplinary Optimization, 50;2014: 1165–1174.

[90] Liu N, Yang P, and Wang W. Design of a miniaturized ultra-wideband compound spiral antenna. In: 2013 IEEE International Conference on Microwave Technology Computational Electromagnetics; 2013:255–258.

[91] Hutchinson BC. Design of an Ultra-Wideband Spiral Antenna for Ground-Penetrating Microwave Impulse Radar Applications; 2015. https://digitalcommons.calpoly.edu/theses/1445.

[92] Wang Z and Xie X. Design and optimization of the antenna applied for detecting the voids in tunnel lining by GPR. In: 2012 14th International Conference on Ground Penetrating Radar (GPR); 2012:131–136.

[93] Jamali AA and Marklein R. Design and optimization of ultra-wideband TEM horn antennas for GPR applications. In: 2011 XXXth URSI General Assembly and Scientific Symposium; 2011:1–4.

[94] Faraji HS, Moini R, Sadeghi SHH, *et al.* Design of a new wire bow-tie antenna for ultrawide-band GPR applications using multi-objective genetic algorithm. In: 2009 13th International Symposium on Antenna Technology and Applied Electromagnetics and the Canadian Radio Science Meeting; 2009:1–4.

[95] Deb K. Multi-Objective Optimization Using Evolutionary Algorithms. Wiley Interscience Series in Systems and Optimization. Wiley, Hoboken, NJ; 2001.

[96] Bechikh S, Datta R, and Gupta A. Recent Advances in Evolutionary Multi-objective Optimization. Adaptation, Learning, and Optimization. Springer International Publishing, Cham, Switzerland; 2016.

[97] Coelho ACC, Lamont GB and van Veldhulzen DA. Evolutionary Algorithms for Solving Multi-Objective Problems (Genetic Algorithms and Evolutionary Computation). Springer, New York, NY; 2002.

[98] Avila S, Krähenbühl L and Sareni B. A Multi-Niching Multi-Objective Genetic Algorithm for Solving Complex Multimodal Problems. In: OIPE 2006, The 9th Workshop on Optimization and Inverse Problems in Electromagnetics, September 13–15, 2006, Sorrento, Italy.

[99] Millard SG, Shaari A and Bungey JH. Field pattern characteristics of GPR antennas. NDT&E International, 35;2002:473–482.

[100] Halabe UB. Condition assessment of reinforced concrete structures using electromagnetic waves, PhD thesis, Massachusetts Institute of Technology, USA; 1990.

Chapter 4
The ground penetrating radar system

4.1 Introduction

Radar is an acronym for RAdio Detection And Ranging. Its operating principle is analogous to the broader pulse-echo (or pitch-catch) technique, except that pulsed electromagnetic (EM) waves (short radio waves or microwaves) are used instead of stress waves. Radar was developed during World War II to provide early warning of approaching enemy aircraft. From there, it evolved into indispensable tools in air-traffic control, shipping, weather prediction, and a multitude of other applications that range from probing distant planets to law enforcement.

During the 1960s, the need for a device to locate nonmetallic land-mines* provided the impetus for the development of high-resolution ground probing radar. Beyond the initial task of mine detection, which is still one of the most important and useful applications of ground penetrating radar (GPR), it is increasingly being recognized as an effective tool in gathering information in a variety of applications. It can be used in virtually any nondestructive assessment of any structure since GPR systems can be configured in ways that take into account specific needs. The system configuration directly affects the performance of the non-destructive testing (NDT) assessment. Since some prior information about the structure under test and the target characteristics is often available (or can be estimated), the system definition plays a major role in the successful execution of assessments.

The application of radar is based on the use of sensors (antennas) that transmit and receive EM waves. First, an EM wave is generated in the instrument's circuitry and propagated to the antenna through a transmission line. This wave is then launched by an antenna in some medium (in general free space but in some cases may be coupled directly into a medium such as soil). The wave propagates within the medium (with attenuation) until it reaches a target with different electrical characteristics than the surrounding medium. Some of the power in the wave is scattered by the target, propagates in various directions, including in the direction of the antenna, and some of it is received by the antenna (either the same antenna in an a-static (also called mono-static) radar, a separate antenna in bistatic radar or an array of antennas in multistatic radar). This received waveform is then used to determine useful information about

*Over 100 million antipersonnel mines are thought to be buried in over 70 countries. Some 24,000 people are killed or injured every year by these antipersonnel mines.

the target such as its presence, electrical properties, dimensional characteristics, and range or location. In free space, the range of radar is only limited by the spread of power, the latter depending on the radiation pattern (or the gain) of the antenna. Under ideal conditions (no attenuation and no spread), the range in pulsed radar only depends on the width of the pulse transmitted by the antenna, that is, by the time available until the antenna transmits the next pulse and may be written as

$$R = \frac{v_p t}{2} \tag{4.1}$$

where v_p is the speed of propagation of the EM wave and t is the time necessary for the wave to propagate from the transmitting antenna to the target and back to the receiving antenna.

In practice, of course, media are lossy and in the case of GPR, they are often highly lossy, limiting the range considerably. In addition, the spread of the wave, which is often described through the gain of the antenna, plays an important role as does the sensitivity of the receiving circuitry. A common approach used to estimate the range of radar is the radar equation (see Section 2.5.2), which does so in terms of radiated and received power from the target as follows [1]:

$$P_R = \frac{P_T G}{4\pi R^2} \times \frac{\sigma}{4\pi R^2} \times A_e \tag{4.2}$$

where P_T is the peak-transmitted power radiated by an antenna with gain G given over the area of a sphere of radius R. The radiated power reaches a target with radar cross-section σ that scatters the incident field and the latter is detected by the receiving antenna with an effective aperture or effective area denoted by A_e. Effective aperture was defined in Section 2.5.2 as the ratio of power received by an antenna divided by the time-averaged power density at the location of the antenna and equals $A_e = G\lambda^2/4\pi$ [m^2] for any antenna either in transmit or receive mode (G is antenna gain). The received power is P_R. For the purpose of this discussion, it is assumed that the transmitter and receiver use the same antenna. The maximum radar range is obtained by substituting $G\lambda^2/4\pi$ for A_e in 4.2 and defining the minimum detectable power, denoted here as the minimum detectable signal $S_{\min}$.

$$R_{\max} = \left(\frac{P_t G^2 \lambda^2 \sigma}{(4\pi)^3 S_{\min}}\right)^{1/4} \tag{4.3}$$

This simple estimate does not consider noise. The equation can be modified to account for noise and defining a minimum detectable signal power $S_{\min}$ at the receiver in the presence of noise as [1]

$$S_{\min} = KT_o BF_n\,(SNR) \tag{4.4}$$

Table 4.1 Radar parameters

Description	**Units**
Frequency	Hz
Peak power	kW
Antenna gain	dB
Target cross-section	m^2
Effective temperature	K
Bandwidth	Hz
Noise figure	dB
Radar losses	dB
Range	km
SNR	dB

where $K = 1.38 \times 10^{-23}$ [J/K] is the Boltzman constant, T_o is the standard temperature of 290K, B is the receiver bandwidth, F_n is the noise figure, and SNR is the signal-to-noise ratio at the receiver. To account for noise, (4.3) becomes

$$R_{\max} = \left(\frac{P_t G^2 \lambda^2 \sigma}{(4\pi)^3 K T_o B F_n (SNR)} \right)^{1/4} \tag{4.5}$$

Table 4.1 summarizes the variables and the units in the radar range equation.

In addition to range, for a moving target, it is possible to determine its velocity using the Doppler frequency shift of the received signal. The spatial direction of the target can be determined using antennas with proper radiation patterns, that, again, are related to directivity of antennas. Equation (4.5) is important not in terms of its accuracy—it assumes lossless media—but rather in the relation between its various components. It clearly indicates a number of trends that must be used as guides in design and specification of equipment as well as some of the conflicting terms. Clearly a higher radiated power is useful but so are the directivities (or, alternatively, the gains) of antennas and the size of targets one looks for. On the other hand, a typically higher bandwidth is associated with sensitivity that is tempered by noise, which increases with the bandwidth and hence reduces the range of detection. Noise figures and SNRs are critical and everything should be done to reduce these. The overall conclusion from (4.5) is that of a system with short range in which one or a number of parameters can be emphasized or optimized, typically at the expense of others.

4.2 Classification of ground penetrating radars

Pulsed radars were one of the first classes of radars developed for military use [1] and later adapted to air-traffic control. These are based on transmitting and receiving a train of modulated pulses. As such, single antennas can be used to send and receive EM waves whereby the transmitting antenna receives reflected signals after the transmission of the pulse is complete. To improve performance, the waveform can be optimized using a variety of modulation techniques. However, other methods have evolved from these early pulsed radars and currently one can select from a number of types of radars suitable for a variety of applications beyond the military and air-traffic control. The availability of various types of radars calls for some methods of classification to aid in discussion.

There are a number of ways this can be done. Perhaps, the most common distinctions between the various types of radars are based on the waveforms and on the modulation they use since these have profound effects on performance and their uses. Another distinction can be made on the basis of frequency of operation. Although there are other classifications (such as based on bandwidth, power, or application), the discussion here is limited to these three classifiers since they are the most critical in terms of system specification and have the most profound effect on performance. Nevertheless, issues of bandwidth, size, and antenna type, among others enter into considerations.

The most common GPRs based on the waveforms they transmit are as follows:

- Continuous-wave (CW) radar. These use continuous sinusoidal waveforms as these simplify circuitry, reduce power, and allow for additional parameters such as the use of the Doppler effect to detect and classify moving targets. There are of course limitations, not the least being the need to restore a reference between transmitted and received signals that overlap in time, and the narrow bandwidth associated with CWs. Some of these issues are resolved through modulation techniques.
- Pulse compression radar. These attempt to use long pulse waveforms and modulations to improve bandwidth. The goal is to obtain performance similar to short pulse radars. Pulse compression radars are popular because the various modulation techniques offer a variety of improvements not possible with other method.
- Synthetic aperture radar (SAR). Inspection of (4.5) shows that one of the important issues is gain or directivity (which, in turn, are associated with effective area and antenna type and size). One way to improve performance is to synthesize the effect of a large antenna through signals from small antennas from a number of physical locations thus increasing resolution. Although SAR is more complex than other radars and requires motion over the target, it is well suited for GPR as it uses the same types of antennas, and the individual signals maintain properties such as phase coherence. The main difference between SAR and other radars is in the imaging section of the system.
- Pulse radar. The antenna transmits pulses that are much longer than the period at the center frequency of the radar. At the end of each pulse, the antenna is ready to receive the reflected pulses. Although pulse radars are some of the simplest conceptually, with relatively simple components, they are limited in a number of ways, not the least in resolution because of the limited bandwidth associated with long pulses.

- Impulse radar. These were developed to emit very short pulses (often of the same order of magnitude as the period at the center frequency) with high peak power (but low average power) and, as a consequence of the short pulses, large bandwidth. For this reason, the technique is often called short-pulse radar, impulse radar, or ultra-wideband (UWB) radar. The method offers high resolution with low power densities and therefore lower clutter.

Radar can be, and often, is classified by its operating frequency. The EM spectrum is divided into bands for easy reference and for allocation to various services (see Section 2.3). Radar manufacturers are constrained by regulations to use specific segments of the EM spectrum by national regulatory agencies. Classification based on frequency band in which they operate is therefore a rather natural classification. Table 4.2 lists the various frequency ranges corresponding to bands in the frequency range from 3 MHz to 110 GHz. Saying that a radar is an S-band radar means it operates at a frequency between 2 and 4 GHz. Note, however, that GPR is restricted to frequencies below 10.6 GHz with specific limits on power and further limits on impulse radar center frequencies [2–4] (see also Section 3.5.1.4). Thus, GPRs are classified in the HF through X bands. Table 1.3 lists a variety of GPRs from different manufacturers with operating frequencies between 20 MHz and 3 GHz. Of course, outside the domain of GPRs, other radars can operate at higher frequencies subject to allocation by legislation and following specific limitations and standards.

The third major method of classification is based on the method of modulation. The simplest method of generation of a waveform is to start with a CW and modulate it to obtain a wave of required properties. The simplest way to think about this is in terms of a single frequency pure sinusoidal oscillator. This can now be formed into a signal with a given bandwidth by switching it on and off to generate pulses, by superimposing a low-frequency waveform over it or by shifting the frequency in some way. For GPR, there are a number of methods of modulation that are useful (see Section 1.3.4 and Figure 1.6). The most common are as follows:

- Amplitude modulation. The simplest forms of modulation are those based on amplitude. This can take the form of classical modulation using a low-frequency waveform superimposed on a CW as is done in communication but in radar it usually takes the form of the on/off modulation whereby the CW is switched on and off at given rates. The generation of pulses in pulse radar is of this type.

Table 4.2 Frequency band designation

Band	HF	VHF	UHF	L
Frequency range	3–30 MHz	30–300 MHz	0.3–1 GHz	1–2 GHz
Band	S	C	X	Ku
Frequency range (GHz)	2–4	4–8	8–12	12–18
Band	K	Ka	V	W
Frequency range (GHz)	18–27	27–40	40–75	75–110

In GPR, this can be done at a fixed rate or variable rate as appropriate for the inspection. This type of modulation is most often used in CW radars since purely (unmodulated) CW radars are not very practical. The modulation is also associated with time-domain GPRs—that is, amplitude modulation generates a time-domain waveform with a bandwidth that can be adjusted by changing the on/off rate. Similar methods can be used in impulse radars.

- Frequency modulation. Again starting with a CW form, it is possible to frequency modulate the signal by increasing or decreasing the frequency of oscillation in a range that allows a sufficient bandwidth for the application. The purpose is to allow the wave to explore the frequency response of the target over the bandwidth and thereby improving response. As with any option one employs, this type of modulation comes with problems, primarily in terms of interference. Nevertheless, frequency modulation is common for radars that operate in the frequency-domain and are generally called frequency-modulated CW (FMCW) radars.
- Stepped-frequency modulation. An improved form of frequency modulation for GPR in the frequency-domain is the stepped-frequency CW (SFCW) modulation. In its principle, it consists of scanning over a range of frequencies (the bandwidth) in discrete frequency steps. The method is simpler and more adaptive than FMCW and has additional advantages such as the ability to measure the phase difference between transmitted and received signals at each frequency and also simplifies post-processing of signals.
- Gated stepped-frequency modulation. A modification of the SFCW method, the gated SFCW modulation adds a short gating period at the start of each frequency step, thereby removing the very strong, early reflections from the received signal and thereby increasing sensitivity to target data.
- Noise modulation. The noise-modulated CW method typically uses a random noise function for modulation and as such improves correlation between transmitted and received signals. In some cases Gaussian noise profiles are used [5].
- Interrupted frequency modulation. In this method, the transmitted frequency-modulated CW is interrupted for a period of time before receiver sampling begins primarily as a means of reducing interference [6].

4.3 Requirements from ground penetrating radar

Depending on the application, GPR can perform one or more of the following tasks:

1. examination of given volumes in order to detect potential targets;
2. imaging of targets; and
3. classification of the characteristics (electrical and dimensional) of targets.

Like any radar system, GPR can be produced in various configurations. The definition of a configuration is related to the performance desired for a given class of applications. In order to determine the performance of a specific configuration, there are some important criteria that should be evaluated (in addition to issues of cost, radiation safety, power, etc.):

1. reliability of detection;
2. accuracy of prediction in terms of location and features of the target;
3. resolution in the entire survey;
4. size and weight of the equipment;
5. materials and properties expected;
6. coupling to the medium; and
7. susceptibility to environmental noise.

The reliability of detection must take into consideration the expected location and characteristics of targets and the medium they are embedded in. Prior information about the degree of non-homogeneity in the surrounding medium should be considered due to the fact that the characteristics of the host medium could be in a range of values, some of which may be close in values to those of the target (e.g., personnel and landmines are designed with dielectric properties to resemble soil). One must evaluate the possible impact of reflections from the surface and other sources of clutter that may exist. Realistic expectations should be part of the evaluation. Similarly, any source of possible false- positives should be carefully evaluated.

Accuracy of target detection and classification is important in most cases and literally critical in others. This includes information about its position, size, shape, orientation, and electrical properties. The importance of accuracy in mine detection was emphasized earlier but the same applies to location of flaws in concrete or position and type of pipes underground. As stated earlier, the host medium and target must have different properties for the incident wave to generate a strongly scattered wave to be detected by the receiving antenna. In addition, the difficulties encountered in resolving any ambiguities in detection should be considered as part of the design.

In GPR surveys, targets must be resolved from each other by one or more spatial coordinates. This is called spatial resolution and measures the ability of the system to distinguish (or discriminate) reflections from different but closely located targets, or the target reflection from surface reflections (clutter) that may interfere with reflections from the target.

Multipath interference or nondeliberate electronic interference can diminish the value or preclude the use of GPR. To mitigate interference, GPR systems should be minimally susceptible (have high immunity) to environmental radio-frequency (RF) interference. The immunity measures the ability of the GPR to perform its functions in the presence of noise. In addition to RF interference (RFI), radar receivers are also affected by the following sources of noise [7]:

1. thermal emissions of the target scene;
2. random currents in any and all components, including semiconductor shot noise;
3. data quantization effects; and
4. purposeful random dithering signals.

Noise can originate from both intentional and unintentional radiators. This is the classical electromagnetic compatibility (EMC) issue caused by external or internal sources. Noise can be broadband or narrowband and may be frequency dependent [7]. However, in GPR, noise can also be due to anisotropy or due to small distributed targets that are not part of the survey such as rocks and roots.

In an assessment, the RF power generated in the GPR unit follows a path through different components and media in the instrument, each with its own transfer function that causes changes in the original waveform. Changes in the waveform also occur as the pulse propagates through various media after transmission. These modifications impact the GPR performance and are important considerations in system specification. Given these effects, the waveform definition is a crucial factor for the specification of GPR systems for NDT, once the targets and host medium characteristics are defined.

Some of the aspects of waveform definition that should be considered are as follows:

1. SNR;
2. avoidance of false-positive detection;
3. resolution required; and
4. EMC issues.

It is important to note that these considerations are mutually conflicting in most applications and trade-offs are almost always resorted to.

As mentioned earlier, GPR systems may use pulsed or CW forms. Most of the commercial systems are based on pulsed waveforms, but CW form systems also exist and have certain advantages. The choice of the waveform and its bandwidth is directly related to the resolution required in the assessment. The waveform bandwidth can be controlled using various types of modulation.

Once the waveform is generated, it must be radiated by the transmitting antenna into the structure under test at the appropriate power level. The antenna must meet the spectrum features of the selected waveform. The available access to the structure is also an important issue in determining the antenna characteristics. In addition, prior information about the test scenario is important, as it facilitates the selection of an antenna structure that is less susceptible to undesired reflections and interference. The antenna structure can impact the GPR readings imposing practical trade-offs. Available commercial equipment show a vast range of antennas of differing sizes and radiation patterns. These characteristics directly impact the signals received from the structure. Significant in this respect is the selection of spacings between measurements which, in turn, are related to the characteristics of the medium under test. The choice of the spatial interval of measurement can improve the readings but it may impact the time spent for the assessment. A careful choice can avoid data volume exaggeration or a spatially aliased response.

The distance between antennas in bistatic, multistatic, and array antennas should also be considered in the system specification because of mutual coupling that necessarily occurs. This feature is related to the radiation pattern of the antenna and its coupling with the host medium (see Chapter 3) and can have detrimental effects on the waveform radiated and received.

Another important issue in GPR systems is the data processing and interpretation. Whereas an experienced GPR user can interpret raw data from measurements without the need for additional data processing, most users need the assistance of GPR software. GPR software tools employ signal processing techniques to improve the measurement and the likelihood of detection. In the remainder of this chapter, we will discuss the important issues of GPR system specifications and some of the signal processing techniques in use.

4.4 System specification

GPRs can be configured for specific applications by simply selecting appropriate components and operating parameters. Some of these can be selected in software, whereas others involve use of hardware modules that may be installed in the manufacturing process or may be user-installed options (such as antennas). Since the system configuration affects the performance of the NDT assessment, any known information about the structure under test and the target characteristics should be incorporated in the system specification. The first and most important choice to be made is the frequency of the survey. This selection affects the primary features of the survey such as depth of penetration and resolution. As can be seen from (4.5), the higher the frequency the shallower the depth at which targets can be detected reliably. Similarly, higher frequencies also mean lower horizontal resolution because the shorter wavelength results in a smaller antenna footprint [8] as can be seen from (1.12) (see also Figure 3.11). The issues of sampling and digitization were discussed briefly in Section 1.3.4. It is also important to take the sampling criteria in space and time into account in relation to frequency selection. This is controlled by the acquisition parameters of the equipment, such as the sampling time and the space between readings. These and additional aspects of sampling will be discussed in Section 4.7 to follow in this chapter. For now, however, to avoid aliasing, the Nyquist theorem provides maximum time and spatial intervals for a given frequency f as

$$\Delta t \leq \frac{1}{2f} \tag{4.6}$$

$$\Delta x \leq \frac{\lambda}{2} \tag{4.7}$$

In terms of vertical resolution (i.e., the ability to discriminate stacked targets), the bandwidth of the system is most relevant and is tied to the radiated waveform. The waveform is another choice to be made. As was mentioned previously, any EM pulse following a path through different media undergoes changes in its original waveform. These modifications will impact the GPR performance, and the waveform definition is a crucial factor with far reaching effects on detection and processing of data.

The waveform, after proper amplification, is transmitted into the host medium. However, there are various aspects that can distort the waveform and modify its spectrum. As discussed in Chapter 3, the antenna is one of the possible sources of distortion of the transmitted waveform. It is therefore important to consider the relevant parameters of the antenna that can generate noise and interference. In general, microstrip antennas are selected since they interfere less due to their nondispersive properties. For these reasons, planar antennas are seldom used in ground-coupled surveys. In air-couple systems, horn antennas are good choices since they can radiate the waveform with fewer distortions over a wide frequency range and at high gains, but their appeal is limited by their size and weight.

The antenna can be viewed as the core of any radar and impacts all other components of the system. In commercial equipment, it is not unusual to find different types of antennas depending on applications and suppliers of equipment. In addition, if warranted, the user can select a dual-frequency system that can operate at a low frequency (with an antenna optimized for the frequency range) for deep surveys and switch to high frequency (and an appropriate antenna) for higher resolution, shallower surveys. Another important characteristic of the antenna is its size. There are applications that cannot accommodate large antennas (e.g., in some civil engineering applications). Equipment notwithstanding, the user can have a significant impact on the success of an assessment by proper attention to important issues such as the spatial interval of measurement. Although this may seem as a minor consideration, improper choices are very difficult to correct in the post-processing stage and may lead to a redo of the survey.

Therefore, the requirements of a GPR system describe the necessary functions and features of the system that need to be defined and implemented in order to extract relevant data that can generate useful information. The following organizes the requirements hierarchically. Based on the survey characteristics, the requirements set constraints and goals in the choice of system parameters. It is important to note that the requirements for a GPR system are always trade-offs between performance and cost. In addition, as will be shown in Chapter 6, post-processing support tools play an important role in eliminating sources of errors in the measurement.

4.5 System requirements

The requirements needed to specify a GPR system are directly related to (4.5) from which it is possible to infer a number of parameters and how they influence the system performance. Figure 4.1 shows a typical block diagram of a GPR system, consisting of transmitting and receiving antennas, a signal generator (SGT), digital-to-analog converter (DAC), RF amplifier (AMP), low-noise amplifier (LNA), analog-to-digital converter (ADC), and digital signal processing (DSP) units.

Figure 4.1 illustrates the flow of the signal in the system. First, the signal is generated with the required waveform modulation. The discrete signal $x[n]$ is then converted to an analog signal $x(t)$ and amplified before being propagated to transmitting antenna TX through a transmission line. The signal then is radiated into air or directly into the host medium (for ground-coupled antennas) to interact with the host medium and targets (these are indicated in the diagram by a generic transfer function $h(t)$). The reflected signal, if any, propagates back to the receiving antenna. This signal is attenuated due to, primarily, propagation losses, to levels that can be

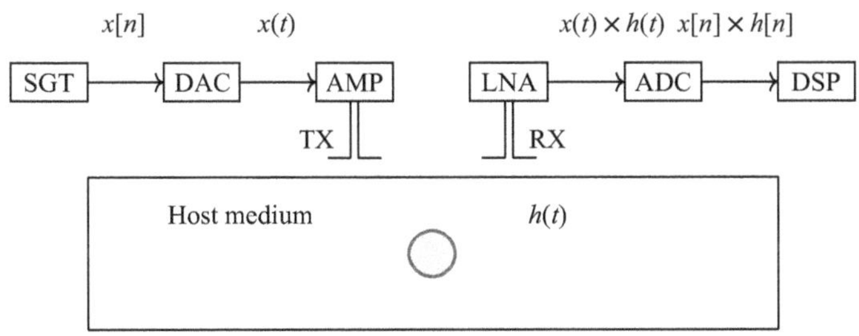

Figure 4.1 Block diagram of a typical GPR

close to the noise floor. To improve the convoluted signal $x(t) \times h(t)$ without adding more noise, an LNA is used before the signal is handled (displayed, processed, etc.). To be processed by a computer, the analog signal is then converted to a digital signal ($x[n] \times h[n]$) using sampling and quantization algorithms. The following sections describe the various components of the system in some detail.

4.5.1 Signal generator

The waveform defines the type of GPR system. According to (4.5), it also affects other parameters since it defines important quantities such as operating frequency and the bandwidth of the system, including that of antennas. The signal generator generates that waveform at a low-level amplitude and power. As stated before, there are various types of waveforms that are used in practice and these are typically generated through the use of programmable digital SGTs. In general, the SGT cannot deliver the wave at proper amplitude levels required for propagation in the host medium. A power amplifier is therefore necessary to boost power. At the RFs used in the system, an important issue when connecting the various blocks in the system is impedance matching, which has to be maintained throughout. The short wavelength characteristic of radar requires the use of transmission lines such as coaxial cables to guide the waves. If an impedance mismatch between the signal generator, the transmission lines, and the amplifier exists, there will be reflected waves and a standing wave will be generated distorting the waveform and wasting power in the system. In many cases, SGTs as well as other components are modular allowing for adaptation to various applications. The type of SGT used depends on frequency range and application. At relatively low frequencies, the signal may be generated by simple oscillators. At higher frequencies, harmonic oscillators, in which a signal rich in harmonics is generated and an appropriate frequency is selected through filtering, may be a better choice.

4.5.1.1 GPR waveform types

The GPR waveform may vary according to the needs of specific applications or classes of applications. There is no single waveform that can be used for all applications, since each has advantages and drawbacks depending on the usage. The choice of a waveform requires basic knowledge of the system, host media, and target characteristics. There is always a compromise between desired objectives such as resolution and depth of penetration in the host medium that must be addressed. Given these requirements, the design of a proper waveform with optimal bandwidth and center frequency can be a difficult undertaking yet, the selection of an appropriate waveform has far reaching consequences on the performance of the radar and its design [9–13].

The basic distinction between the various possible waveforms is between those appropriate for time-domain (pulsed) and frequency-domain radars, with pulsed radars being more common. Starting with time-domain waveforms, which use amplitude modulation, there are various possibilities available to the designer. The most common choices are Gaussian pulses and pulsed derived from Gaussian forms. Figure 4.2 shows a typical Gaussian wavelet [14,15].

This type of waveform can be used in GPR assessments but, because Gaussian wavelets have a DC component, a considerable amount of power is lost. In addition, when the Gaussian waveform and its first derivative are coupled into media under test, it becomes difficult to identify the pulse from its reflected component. To alleviate these difficulties, a modified waveform, derived from the Gaussian pulse, may be used. This is called the Ricker wavelet [16,17].

The Ricker wavelet is the second derivative of the Gaussian pulse and is symmetric in the time-domain but not in the frequency-domain. This wavelet is often used in problems requiring low-frequency content (narrow bandwidth) of the pulse. The Ricker wavelet is as follows:

$$x_{dg} = A\left[1 - \frac{2}{\tau^2}(t - t_0)\right]\exp\left(\frac{(t - t_0)^2}{\tau^2}\right) \tag{4.8}$$

where A is the amplitude, t_0 is the time of pulse occurrence, and τ is a time constant. Figure 4.3 illustrates the Ricker wavelet.

The center frequency is at the geometric center of the pulse bandwidth. As stated previously, the frequency spectrum is important because it defines the resolution

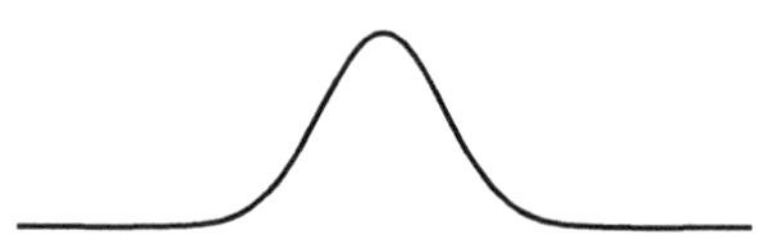

Figure 4.2 Gaussian wavelet (vertical axis represents voltage, horizontal axis represents time)

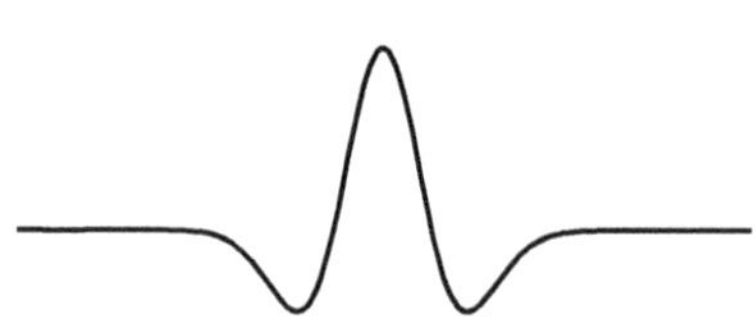

Figure 4.3 Ricker wavelet (vertical axis represents voltage, horizontal axis represents time)

and the depth of assessment. However, the received waveform might differ from the transmitted Ricker wavelet due to, for example, the antenna transfer function.

A common method of amplitude modulation consists of first generating a purely sinusoidal signal at constant frequency and amplitude. Modulation is performed by changing the amplitude from one to zero or zero to one (on or off) at different times during the inspection. Figure 4.4 illustrates the CW amplitude modulation signal for GPR also known as a pulsed system.

Although amplitude modulation is more common, frequency modulation is also possible and is becoming popular with GPR manufacturers because the resulting waveforms are more robust to noise. Frequency-domain systems can also use a number of different types of waveforms through the use of modulation (see Section 1.3.4 and the introduction to this chapter). In the FMCW system, the frequency of the waveform is linearly increased in a defined bandwidth during a given time T (Figure 4.5). Frequency-domain systems also require less power than time-domain systems. One significant disadvantage, however, is coupling between the transmitting and receiving antennas.

There is of course no specific rule as to what waveform to use, how to generate it, or how to select it. A proper procedure that can be used to select the waveform parameters is as follows [18]:

1. predict the gain parameter;
2. calculate the spreading/attenuation losses that can be tolerated;
3. determine the maximum penetration depth in wavelengths;
4. calculate the upper frequency bound f_U of the bandwidth;
5. express the required resolution as a number of wavelengths;
6. determine the fractional bandwidth B/f_c of the system; and
7. calculate the bandwidth B and center frequency f_c.

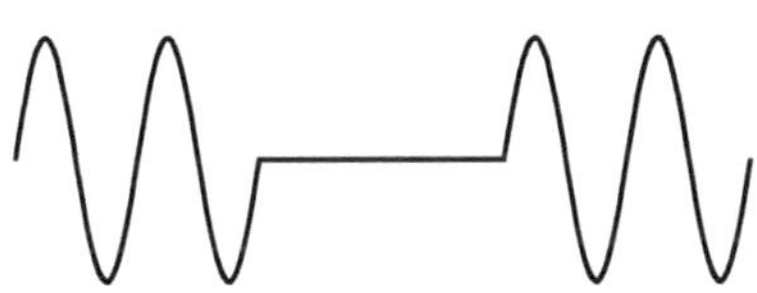

Figure 4.4 Pulsed continuous amplitude waveform (vertical axis represents voltage, horizontal axis represents time)

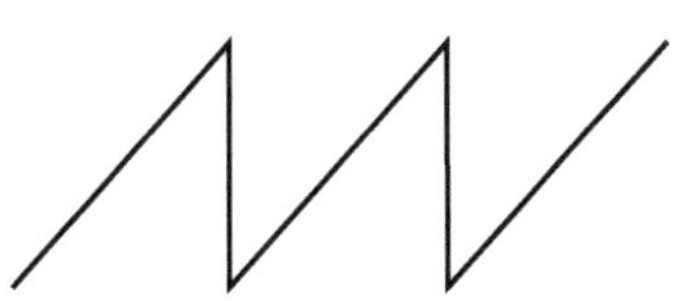

Figure 4.5 Frequency (vertical axis) over time (horizontal axis) for an FMCW form. In this case, frequency increases linearly for a time T and then repeats indefinitely

Clearly not all of these quantities are known and some rely on the user's judgment or experience. Nevertheless, the procedure allows at the very least, an approximate method of selection.

4.5.2 Bandwidth

The compromise between resolution and depth of penetration is a crucial issue for GPR detection and evaluation. Whereas the depth of penetration is related to power (as well as frequency), resolution is directly related to the bandwidth. However, it is important to recognize that the host medium acts as a low-pass filter that modifies the transmitted spectrum in accordance with the electrical properties of the medium in which the waves propagate. In this context, it is the bandwidth (B) of the received signal that is important, as shown in (4.9), rather than that of the transmitted wavelet. Given the bandwidth, the depth resolution (ΔR) can be written as

$$\Delta R = \frac{c}{2B\sqrt{\varepsilon_r}} \tag{4.9}$$

As discussed previously, the higher the bandwidth the better is the GPR capacity to resolve targets as (4.9) clearly indicates. However, as the receiver bandwidth increases, so does the noise power and, as a consequence, the SNR deteriorates (see (4.4)). As a rule, for high SNRs, it is desirable to use longer pulses, which means a narrow bandwidth. Bandwidth also affects the other blocks of the system. For instance, the amplifier has to provide a constant gain over a specific bandwidth. If it fails to do that, part of the signal will be compromised with nonlinear data. As described in Chapter 3, the antenna design is largely affected by the bandwidth requirements. Antenna parameters such as input impedance and gain can vary drastically with frequency and, as a consequence, introduce distortions in the transmission lines and increase losses in the system.

In order to resolve closely spaced targets, GPR systems require a high bandwidth that, in turn, means dealing with time-varying high-frequency EM waves in the presence of time-varying noise. The high-frequency waves, detected by the receiver in an analog form, need to be converted into digital format for processing and interpretation. The conversion is implemented in two broad steps: sampling and quantization. The first step is particularly difficult since high-frequency signals must be sampled using high sampling rates. This problem can be alleviated by using sequential sampling techniques that can reduce the sampling rate by a factor of 10 or more. However, even with a considerable reduction in the sampling frequency, the ADC needs the samples to be stable in order to collect the exact data. This is achieved by sample and hold circuitry with a short dwell time. The dwell time is the time the received signal is connected to a capacitor to be charged to the voltage of the EM wave received (the sampled value is "held" as a voltage across a capacitor for the time needed by the ADC to read the value). As a consequence of the high frequencies and short sampling times (high sampling rates), ADCs must be able to read data at the same rates. In addition, they are required to have high resolution for high-frequency signals or some of the data may be lost. This necessarily means that ADCs for GPR are complex and expensive components.

Another important specification of ADCs is their dynamic range that is related to the number of bits available and hence to resolution. For instance, a 10-bit ADC has a dynamic range of 60.2 dB. This ADC is capable of resolving three decades of amplitudes from the minimum to the maximum value in the signal (1,024 levels). The lowest value the ADC can resolve is an important property, since this minimum amplitude can be confused with the equipment noise floor. In a situation such as that of landmine detection where the target is designed to resemble the ground, the ADC plays a decisive role since the dynamic range will define how well the GPR can distinguish between materials with closely valued electric properties. The ideal configuration is to have as many bits as possible in order to reduce the quantization noise. However, as the number of bits in the ADC increases, the cost increases. As well, the number of bits that are practical is limited by the available technology.

4.5.3 Amplifier

As discussed in Chapter 2, propagation losses are perhaps the most important obstacles to GPR assessment of targets. In order to overcome this, the signal generated is amplified prior to transmission into the host medium. Power amplifiers are a recurrent component in modern communications. Amplification of the signal occurs in multiple stages (driver amplifiers, high power amplifier). A requirement for the power amplifier is to have low internal noise (due to electronic components, type of amplifier, etc.) and to cause minimum distortions to the input signal (e.g., due to nonlinearity in circuits). Some electronic components such as transistors are inherently nonlinear. These devices introduce distortion in the amplification process. For radar systems, the non-ambiguity function of the waveform needs to be preserved when the waveform is amplified. A nonlinear amplifier can distort the amplified waveform before it is transmitted, and that can cause intermodulation products (and consequently spectral growth), phase distortion, ambiguity in the Doppler effect (if used), and widening of the range sidelobes [19].

There are different applications of GPR ranging from detection of ice and snow layers using airborne radar to handheld systems for the detection of small inclusions in concrete. To cope with this broad range, there are different possible topologies for the amplifier ranging from configurations as simple as an amplifier per antenna or connection of multiple amplifiers in parallel at the cost of increased complexity and lower efficiency. Another choice is between vacuum tube and solid-state amplifiers. Vacuum tubes are used for high-power systems (up to 1 MW), whereas solid-state amplifiers are used for low power (typically up to about 100 W but higher powers are possible). Solid-state amplifiers are less costly, do not need a warm-up time, are smaller and lighter, require simpler power supplies, and are cheaper to operate. In addition, they can be used for UWB applications (up to 50% bandwidth) and exhibit improved reliability as indicated by their increased mean time between failures (MTBFs) figure [1].

In general, power amplifiers are designed to cope with the trade-off between linearity and efficiency. The less power is solicited from the amplifier, the better is its linearity. In order to specify an amplifier for a GPR system, the designer must pay attention to the following parameters:

1. required amplification (or gain);
2. bandwidth of operation;
3. average power as a function of frequency;
4. transmission/reception duty cycle; and
5. MTBF.

In addition, since the power level at the transmitter is high and the receiver is in close physical proximity, the amplifier must be isolated from the receiver to avoid interference.

Although not explicitly shown in Figure 4.1, timing and synchronization functions must be available to coordinate the operations of the transmitter and receiver. For example, these are responsible to turn on the transmitter (i.e., connect it to the antenna), while the receiver is disconnected and disconnect the transmitter when the receiver is on. They also synchronize the ADC, DAC, and LNA units.

4.5.4 Power

Equation (4.5) relates the received and transmitted power in a radar system given the environment experienced by the EM wave. This relation is important in determining the dynamic range of detectable signals. In some applications, the user may want to raise the transmitted power so the amplitude of reflected fields from a target can be distinguished from the pervasive random signal errors due to various sources of noise. Another possible action is to reduce the noise floor in the equipment so the signal can be obtained without distortion at lower levels.

In this respect, the sensitivity of the system is of primary importance because it determines the minimal field level that can be acquired without distortion at the receiver. Range resolution can be improved if the bandwidth of the transmitted signal is increased (or the pulse width reduced). These influence the dynamic range by placing limits on the maximum transmitted amplitude that can be achieved with the GPR front-end amplifier.

Increase in transmitted power is an option that should be undertaken with care. Aside from the possible effects on distortions and noise associated with higher power, the desired power may not be available. In handheld and field devices, the power is often supplied by batteries and the limit on power can be severe. Much more effort should be directed toward improving parameters such as the noise floor, sensitivity, optimal waveforms, and the like, to obtain the sought after performance at minimum power levels.

4.5.5 Antennas

Antennas, as the core component of any GPR system, affect the overall capabilities of the system. The selection of an antenna depends on a number of parameters as discussed previously (see also Chapter 3). Antenna size and antenna integration with other components of the system are important features to be considered. The planar resolution is important in distinguishing targets at the same depth and is significantly affected by the antenna design. In general, to achieve an acceptable planar resolution requires a high-gain antenna. This necessitates an antenna with a large aperture at the lowest frequency transmitted. To be able to use small antenna dimensions and high gain requires the use of high carrier frequencies, which, in turn, may not penetrate the medium to a sufficient depth. When choosing equipment for a particular application, it is necessary to compromise between planar resolution, the size of the antenna, the scope for signal processing, and the ability to penetrate sufficiently deep into the medium under test.

Antennas, in addition to their electrical functions, are in the first line of physical interaction with the testing environment. They are exposed to effects ranging from dust, to humidity, to physical impact. As such, they must be robust physically and often must be protected from the environment. In handheld portable equipment, seemingly secondary properties such as weight become important and some antennas may not be appropriate even though their performance may be superior.

4.5.6 Low-noise amplifier

The LNA is an important component of the GPR system since it is tasked with the amplification of very weak analog signals received from targets, without degrading the SNR. The LNA must match the antenna impedance at the frequency (or frequencies) of the signals to be amplified. Because of the requirement for low noise, one has to also take into consideration the frequency response of the amplifier (bandwidth) as well as the internal sources of noise produced by its own components.

In addition, the maximum and minimum signal amplitudes must be known in order to properly specify the LNA. Although the noise figure is the primary requirement, the LNA must also be highly linear since a major source of distortion in an LNA is caused by nonlinear conversion of input voltages. The LNA typically operates at low power but high-voltage amplification. While it does not have some of the constraints of the transmitter power amplifier, others, such as shielding, isolation, and immunity, to interference (especially from the transmitting antenna) are more critical.

4.6 Data acquisition modes

In the use of GPR for assessment, the way the measurements are taken is another important parameter that the user has control over. The method used for data acquisition can be modified to affect specific outcomes. For example, one may use a specific method to increase sensitivity or to measure properties such as speed of propagation in the host medium. There are a number of modes appropriate for radar data acquisition. They are chosen according to the application. Some of these modes and their common usage are described in the following subsections.

4.6.1 Common offset mode

The common offset mode is a mode often used in GPR surveys, whereby the transmitter and receiver antennas are fixed at a given distance from each other. The two

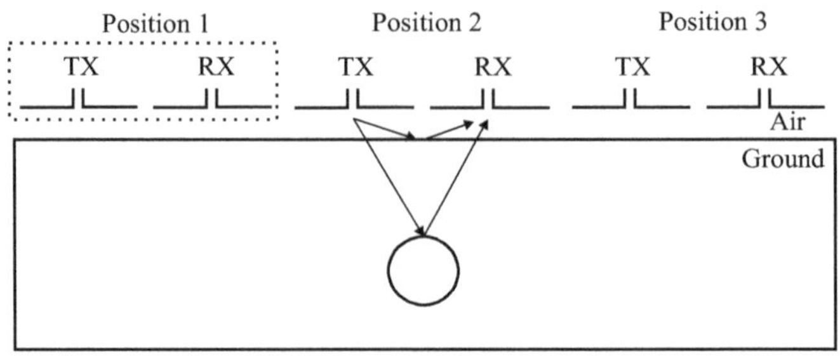

Figure 4.6 Common offset acquisition mode

antennas move in tandem over the surface to be scanned in order to measure travel times of the EM wave.

Fundamentally, this mode is called SAR, whereby the antenna is moving, and measurements taken at different time instances to display the vertical axis corresponding to different antenna positions forming the horizontal axis of the radargram. SAR is frequently used in civilian and military surveillance applications and it has the same purpose in GPR.

One of the problems when using the common offset mode with air-coupled antennas is the reflections from the environment. This error source can diminish the azimuthal or planar resolution. Figure 4.6 depicts the common offset mode in GPR surveys.

4.6.2 Common source and common receiver modes

The common source and common receiver modes are derived from the common offset mode. In these modes, the receiver or the transmitter is kept in a fixed position while its counterpart moves. In the common receiver mode, the operator changes the position of the transmitter in order to create an image with various offsets. Alternatively, the transmitter is kept at a constant location and the receiver moved to affect the same response. These methods are commonly applied in homogeneous materials using planar antennas. Figures 4.7 and 4.8 show how the common source and common receiver modes are applied in GPR surveys.

4.6.3 Common midpoint mode

Another modification is the common midpoint mode (CMP mode) in which one point of acquisition is obtained by the GPR system using different positions of the group transmitter and receiver. In this mode, the transmitter and receiver are moving with increasing or decreasing offsets symmetrically about a defined midpoint. The midpoint may be arbitrary or it may be based on prior knowledge of targets or features.

This method is often used to determine the EM wave velocity in the material under test. This information is useful in determining the electrical properties of the medium

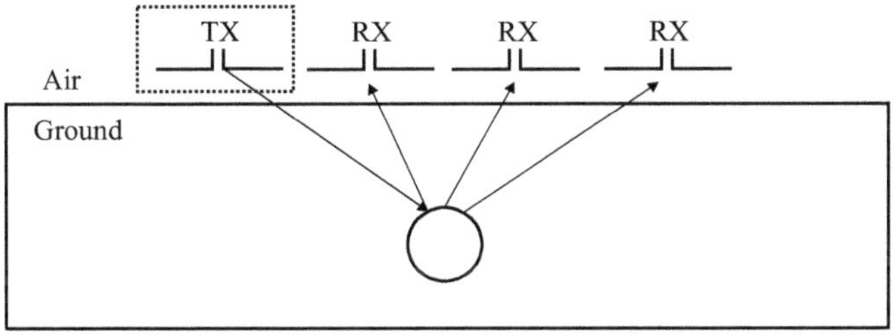

Figure 4.7 *Common source acquisition mode*

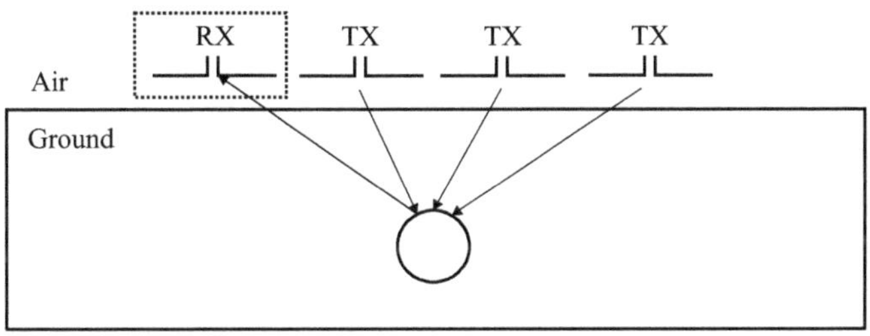

Figure 4.8 *Common receiver acquisition mode*

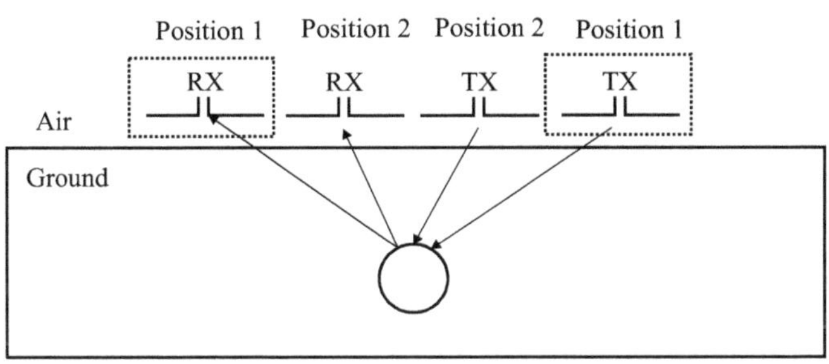

Figure 4.9 *Common midpoint mode*

such as permittivity and conductivity. Figure 4.9 shows the antenna configurations in the CMP data acquisition mode.

4.7 Signal processing

The objective of signal processing applied to GPR assessments is to extract information from raw data and help in creating an image that can be readily interpreted by the operator. Image processing and interpretation procedures separate and classify the reflections into different target categories and then map the relevant reflections to produce interpreted planes, sections, or other diagrams. The goal of these procedures is to select and extract target information from unwanted signals.

In general, the unwanted signals come from early reflections (i.e., before target information reaches the receiving antenna), refraction, or scattering of waves are not related to the target. Reflections may originate in the air-structure interface, structure geometry imperfections at the surface, or improper surface antenna coupling. Early refracted fields can also occur in structures made of heterogeneous materials, in which properties can vary statistically with position. Scattering appears when objects are not related to the target produce signals with a high enough intensity compared to those from real targets. An obvious example is when one is looking for cracks in concrete structures and receives signals from reinforcement steel bars. All these effects (and others) may be bundled in the term "clutter."

Clutter is considered to be one of the primary limitations of the GPR technique. In some applications, the target is buried in shallow depth and the clutter is superimposed on the target response to form a composite signal. Only in specific situations can one remove all the clutter from a GPR reading. If the reflected field from a possible target is small, the signal-to-noise clutter (SNC) ratio may be too poor for the target to be detected. In addition, incoherent random noise and interference from other systems may be present and that can also disturb the readings from the targets.

Signal processing techniques aim at removing signals that interfere with the targets response. In general, the SNC and SNRs can be improved using spatial-, time-, or frequency-domain filtering. However, in some situations, it is possible to apply simpler techniques such as time-varying gain or offset removal. It is important to note that when it comes to signal processing techniques, there are many options available and it is not always obvious which are best. It is clear, however, that there is no single method that can work for all possible GPR specifications and applications. Some of the methods may not be suitable for the specific GPR system or may not be compatible with the resources needed for its implementation. Critical applications such as landmine detection require not only the location but sufficient information is also needed for the safe extraction of the landmine. Over the years, various system specifications and signal processing techniques have been developed specifically for this particular application because of its importance. Other applications have seen their own developments based on their specifications. There are, however, some methods that can be used in most applications of GPR, and some of these are available either integrally with the system or may be available as options from manufacturers.

In this work, the discussion of signal processing techniques is limited to pulsed and stepped-frequency GPRs as these are representative of the variety of GPRs available. Pulsed systems collect data in the time-domain, whereas stepped-frequency systems generate data in the frequency-domain and then convert this information to the time-domain. These data are collected in order to obtain spatial information about the structure under test. This spatial information can be obtained on a line, a plane, or a volume. The basic data are that on a line through the depth of the medium under test and result in a trace called an A-scan. The A-scan is the response obtained by the radar when positioned at a fixed location on the surface of the medium (see Figure 4.10(a)).

Multiple *A*-scans may be obtained by moving the GPR in any one direction on the surface (usually on a well-defined line) to get a structure's side view. The result is a *B*-scan that provides a plane view in depth (i.e., a view over a plane made by the line over which the *A*-scans are obtained and the vertical dimension through the medium) (see Figure 4.10(b)). Sometimes, this composite view of *A*-scans is sufficient but in many application information in three dimensions or on a plane at the depth of the target (i.e., a horizontal plane) is required. To do this, measurements are performed all over the structure surface, usually by scanning over parallel lines to cover the whole surface. This procedure is called a *C*-scan and results in a data set that can produce a 3D view of the structure, shown schematically in Figure 4.10(c). Signal processing techniques can be applied directly to *A*-, *B*-, or *C*-scan.

The *A*-scan provides one dimensional, directional information about the depth of the structure. It contains not only targets information but also undesired signals, including the first reflection from the surface, the receiver/transmitter EM interaction, thermal noise, interferences from other systems, and, in some situations, early reflections from shallow heterogeneities. In almost all cases, the amplitude of the signals reflected from targets is low compared to the undesired signals but, because reflections occur at different depths, they are time separated.

Some of the unwanted components of the signal can be removed using window filtering techniques applied directly to the *A*-scan signal. These typical filters are finite impulse response (FIR) filters. The main advantage in using FIR filtering is the linear phase response, since this is important for EM wave propagation in lossy dielectrics. Early time gating is the simplest method of windowing and consists of eliminating all unwanted signals prior to a given time (T_{int}), which itself depends on the depth of the target. The choice of the time window is a difficult trade-off, since shallow targets may not be detected if T_{int} is too high. Too low a T_{int} results in eliminating only some of the unwanted signals.

Since the EM wave is attenuated with depth in the structure, some GPR manufacturers offer a time-variable gain in their system in an attempt to improve detection of deep targets. In this method, the gain can be adjusted with depth to amplify deeper

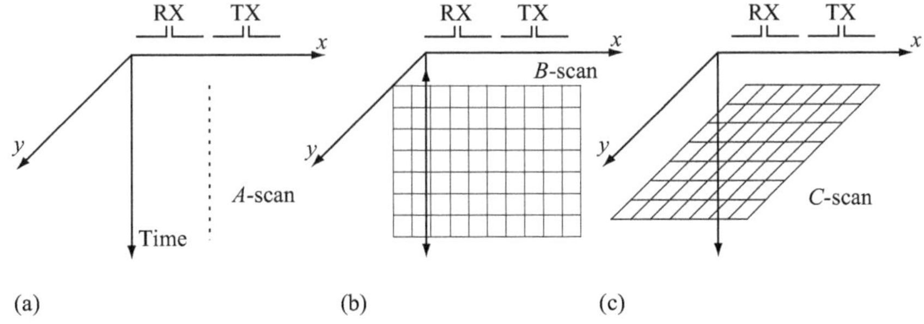

Figure 4.10 Schematic for (a) A-, (b) B-, and (c) C-scan

reflections from possible targets. But, as a result, noise is amplified as well. Similarly, in some commercial GPR options, it is possible to write a customized gain function. However, the material properties are not known accurately in most applications and, as a consequence, the time window definition for the time-varying gain is in fact a major drawback of this technique.

There are other methods that can be applied in the time-domain, including simple mean scan subtraction, background removal, and moving average filtering. Most of these have the same problems as filtering and gain adjustment when it comes to different types of noise. They can also insert errors or amplify noise signals. Frequency-domain techniques have better outcomes since they can subtract sub-bands of interest in the GPR data. As for time-domain techniques, there are many possibilities for frequency-domain processing that can match noise appearing over the whole signal spectrum. One of the most commonly resorted to method by GPR users is a type of high-pass filter also known by dewow. This method is used to control low-frequency interference that can occur from the EM interaction between transmitter and receiver [8].

Noise can also occur at intermediate frequency bands. This can be due to EMC issues, hardware intermodulation, or other transmitters in close proximity operating in the same frequency band. These situations may require the use of notched filtering. Notch filters can be obtained in several ways starting with FIR and infinite impulse response (IIR) filters. IIR filters are less desirable because of their nonlinear phase response, a property that can insert errors when testing a medium with lossy impulse response characteristics.

Another approach for noise removal is to consider the received signal as a convolution of several impulse responses. As discussed earlier, the GPR problem can be viewed as a system where the input signal is a designed waveform that suffers modification as it passes through the many components of the system and media until its arrival at the receiver. The initial signal is convolved with the impulse responses of each component or block, including the antenna impulse response and the medium under test. Deconvolution techniques aim at separating these signals. Deconvolution can occur in spatial-, time-, or frequency-domain and can improve depth resolution in many GPR assessments. Deconvolution is a time-consuming task in the time-domain, whereas in the frequency-domain it is performed by simple multiplications.

Refocusing EM-scattered fields to their true temporal or spatial location is a necessary goal of testing and measurements with GPR. This can be done with deconvolution in the spatial-domain, also known as migration. Migration techniques were first applied to seismic data and then adapted for GPR *B*-scans. Improvements in spatial location for GPR data is often done using time- and frequency-domain migration techniques such as Kirchhoff and frequency wave-number migrations [20–24].

The Kirchhoff migration uses ray theory approximation based on hyperbolic representation of the target's reflected field and is also known as a reverse-time wave equation migration. Frequency wave-number migration or *F–K* migration refocuses the scattered fields using an interpolation of the wave-number after a fast Fourier transform (FFT) of the time-domain data is performed. In homogenous media, where there are no changes in lateral wave velocity, *F–K* algorithms have better lateral resolution than the Kirchhoff algorithm. In addition, F–K algorithms are faster than the Kirchhoff ones due to the use of the FFT. When the surface is uniform and deterministic, both the methods provide proper performance in retrieving the target true location. However, when dealing with rough surfaces media, or statistically defined surfaces, both suffer in performance.

Signal processing techniques are classified on the basis of various criteria. One interesting classification divides the techniques based on their goal. These include SNR improvement, vertical resolution, horizontal resolution, and information extraction for interpretation. These procedures may be performed on data in the space–time-domain (x, y, t), the F–K domain (k_x, k_y, f), and space–frequency-domain (x, y, f).

4.7.1 System abstraction

The GPR may be viewed as a system that has an input and an output and, therefore, given the input and the knowledge of the system, one should be able to obtain the output. In many types of system, this is indeed the case but not in the GPR system since, in the best of cases, we may only have partial information on what lies between the input and output. In others, we may have no information at all on targets and other media. Figure 4.11 shows a GPR scenario that includes, in addition to well-defined transmitting and antenna blocks, the specimen under test and its embedded targets (a more detailed representation of the "system" block in Figure 4.11 can be seen in Figure 4.25). The latter are either not known or only poorly known and hence the transfer function for the assessment cannot be assumed to be fully defined.

Assuming for now that antennas do not interfere with the signals on the input and output allows one to model the input signal by a waveform with desired characteristics, disregarding the effects of the transmitting antenna. The goal is then to analyze the output to determine the presence of targets. This approach is arbitrary and the result cannot be unique. For instance, one could be interested in verifying the wave velocity

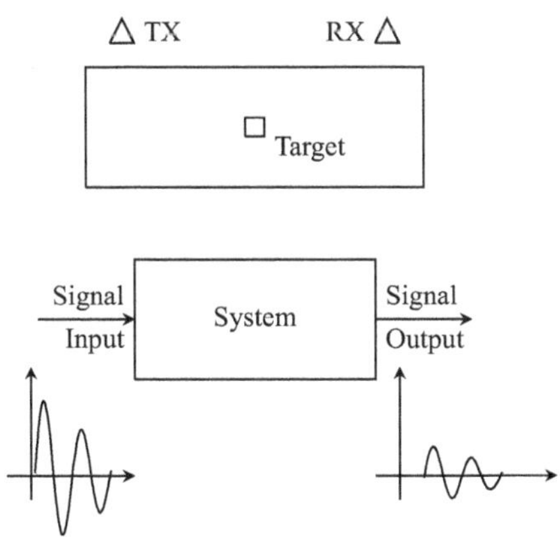

Figure 4.11 Simple representation of a GPR system

or the delay imposed by the system. The analysis follows the only points of interest which are the signals and how to treat them.

In a GPR assessments, Maxwell's equations govern the relations between input and output. They were discussed in Chapter 2 and will not be repeated here. The interest is to see the GPR as a system, neglecting some details such as the effects of antennas on the waveforms and highlighting others such as the influence of targets on the received signal. This simplification is imposed by the very complexity of the system.

Signals from an assessment may be obtained in continuous or discrete time. Whereas the physics of the GPR assessment is defined in continuous time, the signals are processed by digital processors involving computational operations that use discrete data.

4.7.2 Digital signal conversion

In the GPR assessment of structures, signals are transmitted and received as continuous-time signals. Processing of data must, necessarily be done by digital computation, necessitating analog-to-digital conversion as described in Section 4.5.2. There are two basic steps in the conversion, sampling and quantization, both of which have profound effects on the signal.

The receiving antenna detects the signal from the structure as $x(t)$. The signal is sampled at a uniform rate, once every T_s seconds. The continuous signal $x(t)$ now becomes a discrete-time signal $x(nT_s)$. In the next step, each sample of the discrete signal is transformed into its corresponding binary value through the quantization process of ADCs.

The sampling rate is directly related to the sampling frequency that, in turn, impacts the number of samples in the two-way travel time of the EM wave in the structure under test. The precision of the binary representation is directly related to the number of bits used and hence the number of quantization levels (see Section 4.5.2).

Because of the importance of signal conversion in the processing of signals, the foundations of analog-to-digital signal conversion and the main principles underlying the design of ADCs are discussed here in some detail. These concepts can be applied to any application where DSP is required.

Sampling: The discrete form representation of the received signal is fundamental for the target detection and characterization. The representation of a continuous-time signal in discrete form is obtained using the sampling theorem that states that if a signal is band limited and if the samples are collected periodically at a rate related to the highest frequency in the signal, the samples represent the original continuous-time signal accurately. Thus, the sampling process is the initial step for the DSP of the signals scattered by targets.

A signal can be said to represent a pattern that varies with time. Denoting the GPR signal as $x(t)$, it is possible to represent it as a sequence of numbers, $x[n]$. This sequence of numbers can be described as a sequence of samples collected from a continuous-time signal with sampling period T: $x\,(nT_S)$.

Sampling is the most important step for DSP. In a sampling process of a continuous function, the goal is to preserve the information representing a target return signal while still being able to process the data digitally. This is done by recording a sample at constant spaced instants T_s seconds apart and dismissing the information between samples. If the signal is band limited and sampled according to the Nyquist theorem, it can be reconstructed avoiding aliasing. In general, GPR signals are band limited, although high-frequency contributions are less pronounced.

As a result of sampling, we obtain an infinite sequence of samples spaced at sampling interval T_s and denoted as $x(nT_s)$, where n takes on all possible integer values and nT_s are sampling points. From this, one can define the sampling frequency and angular sampling frequency as

$$f_s = 1/T_s, \quad [\text{Hz}]; \qquad \omega_s = 2\pi/T_s \quad [\text{rad/s}] \tag{4.10}$$

The discrete signal can be written as

$$x[n] = x(nT_s) \tag{4.11}$$

The sampling process can be viewed as the multiplication of the continuous-time signal by an impulse train [Equation (4.12)] in the time-domain, which is equivalent to convolution with an impulse train in the frequency-domain. This process generates multiple copies of the signal's original frequency content.

$$x(t) = \sum_{n=-\infty}^{\infty} x(nT_s)\delta(t - nT_s) \tag{4.12}$$

Equation (4.12) is the key to understanding why it is necessary to choose an appropriate sampling frequency. Since the spectrum of the received waveform is replicated over the broader spectrum, it is necessary to choose the distance between the copies. This is done on the basis of the Nyquist theorem to avoid aliasing. Aliasing is a common problem that occurs in the frequency content of a sampled waveform performed with the wrong sampling frequency. The signal is aliased when high frequencies appear in the spectrum position of low frequencies. Spatial sampling is also important for GPR resolution. It is defined as the interval in which the waveforms are collected over a surface. There is a trade-off here as well since the ideal option would be the smallest spatial step size possible, which, however, results in more data to be analyzed.

Quantization: Following sampling, the signal has a discrete representation in time. In order for the signal to be acquired and stored, its amplitude must be represented in discrete levels so it can be stored and manipulated digitally. The trade-off here lies in choosing the appropriate number of discrete levels to properly represent amplitudes, something that is done through choice of the number of bits in the ADC. If the number of bits is too small (resulting in a small number of discrete values for amplitudes), there will be noise generated by the difference between discrete quantities and the analog quantity. In addition, poor digital representation tends to cluster samples (i.e., what are different amplitudes in the analog signal, may be represented as the same amplitudes after discretization) resulting in a banded GPR image. On the other hand, too many bits result in a system that is unnecessarily expensive in terms of devices and operations with little to be gained in return.

Time sampling: Since the GPR system is digital, the user has to choose the sampling period of the data received from the target. This is an important parameter with significant effect on data quality. Although there is no exact method of selection, a rule of thumb for geophysical applications indicates that the sampling rate should be six times the center frequency of the antenna response [25]:

$$T_s = \frac{1{,}000}{6f} \tag{4.13}$$

where T_s is the time interval for sampling in microseconds and f is the center frequency in MHz. As with other parameters, the selection of the sampling time is a matter of trade-off. The smaller the sampling rate the greater the noise influence in the data. The user can always resort to the Nyquist theorem, using the highest frequency within the antenna bandwidth at which there is significant power to define the sampling time.

4.7.3 Data processing

Successful interpretation of GPR data depends on the quality of data received. This is relatively easy in simulated configurations where properties are clear and well controlled, and the data are free of most noise components. In real-world problems, however, the variability of sources complicates the issue considerably. The propagation of EM waves in inhomogeneous media, diversity in target geometries, and changes in the assessment configuration may corrupt the data. Given that this is the reality faced by the user, obtaining appropriate data for further analysis by imaging algorithms and solving the inverse problem depends on the following:

- prior knowledge on the materials involved in the assessment and their electrical properties;
- system parameters such as the waveform center frequency and bandwidth;
- information on how the assessment was performed to allow future corrections if necessary (identification, location, unusual circumstances, weather conditions, and the like);
- design a data processing flow based on the characteristics listed earlier; and
- careful selection of the parameters for each data processing step according to the results.

Each user may find the best flow based on the application at hand. Each application may have its particular spatiotemporal patterns that may require different signal processing approaches. In addition, noise and interference can occur in different ways. If the material is lossless and homogeneous and the target characteristics are sufficiently different from the host medium, simple preprocessing techniques (to be discussed next) may be sufficient for detection. Complex scenarios with heterogeneous lossy media require additional steps.

To demonstrate the use of signal processing procedures mostly on data created using numerical models, some of the most common procedures are described next. Although simulated data are convenient for demonstration purposes, some of the methods are demonstrated using real-world data. It is important to note that there is a trade-off when choosing a signal processing technique. Advanced techniques will require more complexity in the system that processes the data but will, in most cases, provide better SNRs. Yet, simple techniques can produce fast results relying on the operator's experience in attributing some key parameters [8]. In view of these considerations, the techniques that follow are presented on the basis of their impact on the raw data and the need for operator experience.

4.7.4 Preprocessing

Preprocessing techniques are procedures applied to the data that restore or improve organization of the data taking into account specific conditions such as start time, velocity, stationary effects, positions recorded, and sampling issues. In this signal processing step, the user is not required to have experience with GPR assessment or some a priori information about the experiment. As a whole, preprocessing procedures are relatively simple and consume little time but are often necessary and can help with steps that follow. The following subsections describe the common methods of preprocessing of radar data.

4.7.4.1 Data editing

The use of GPR to detect and evaluate targets produces large amounts of data. In some circumstances, the readings are not uniformly distributed producing discontinuities in the data. A typical example is the use of air-coupled antennas to verify pavement conditions. For this purpose, it is common to mount the GPR in a vehicle and proceed with the evaluation at a constant speed of the vehicle. In this situation, data can be corrupted or not synchronized with the actual readings due to a variety of effects, including changes in speed, bumps, vibrations, and the like.

In addition, during the assessment, the user may want or need to reposition the equipment. This can generate gaps and errors in the file that must be resolved for proper interpretation. Similarly, the user may want to discard late-time signals to reduce the amount of data to be processed by the signal processing and inverse algorithms. Data editing procedures are used to organize the data by sorting and rearranging considering the actual conditions under which the assessment was performed. As an example, Figure 4.12 shows a simple assessment performed specifically to collect data for this section, data that are used to illustrate some of the concepts discussed here. Note the instrument in the foreground and the antenna unit (on wheels) on the concrete slab being moved by hand. This scenario was used to generate the data used in some of the processing steps discussed here and in the following subsections.

After editing, a two- or three-dimensional image can be generated depending on how data were obtained and the needs of the tests. Figure 4.13 shows a *B*-scan image of the wave progression in the vertical dimension (into the medium). The changes in the wave can be viewed as wave velocity plots or amplitude plots. In this case, the vertical axis of the plot shows the travel time in nanoseconds. It is also possible to display the individual *A*-scans in the two-dimensional image as in Figure 4.14. This

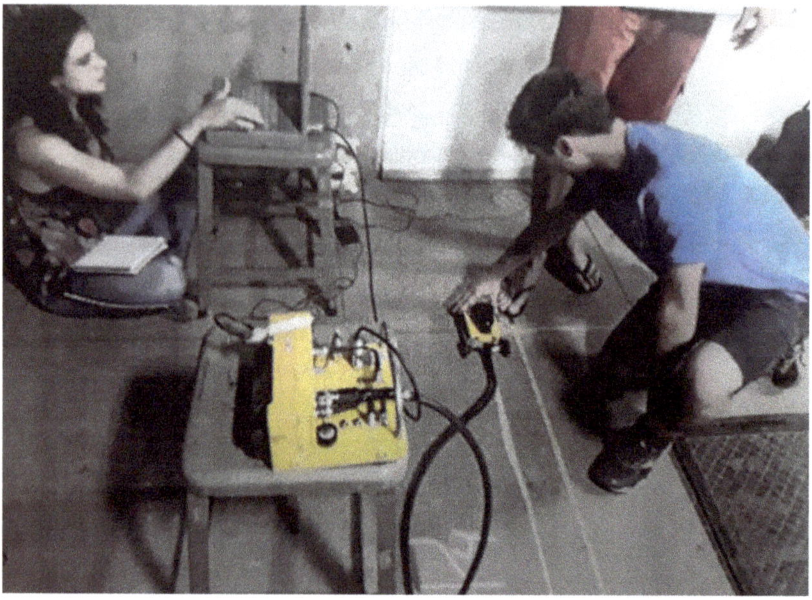

Figure 4.12 GPR assessment in a concrete structure

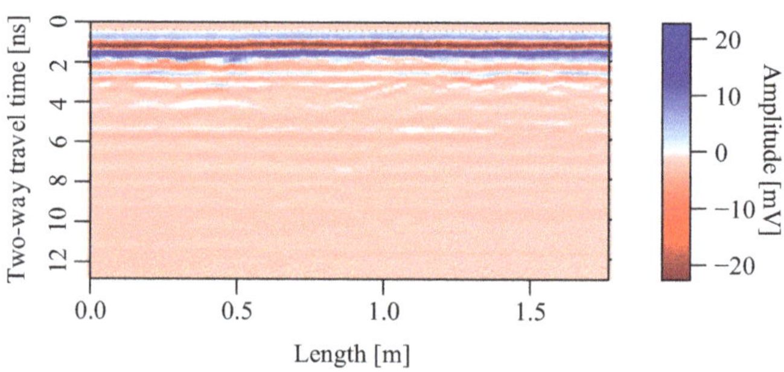

Figure 4.13 B-scan of the assessment in Figure 4.12

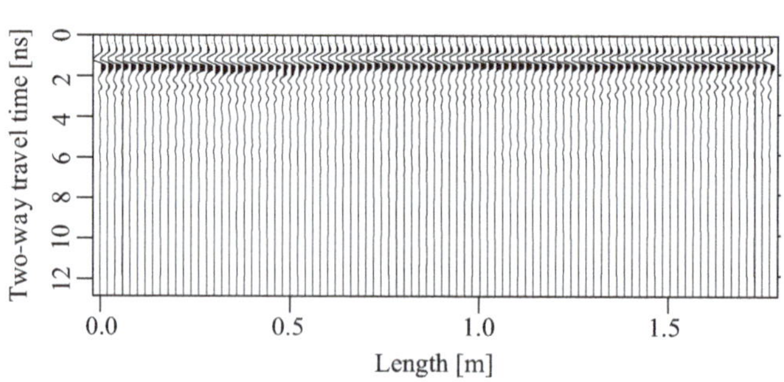

Figure 4.14 Wiggle traces representation of the assessment in Figure 4.12

plot is also called a wiggle trace. It displays individual traces with their amplitudes versus depth.

4.7.4.2 Time zero correction

Viewing Figure 4.13 or 4.14, one observes that the plots start at zero time. The instinctive supposition is that this is the time the signal leaves the transmitting antenna since that is a natural reference point for propagation. However, because targets need to be associated with a depth, the surface over which scanning occurs (ground surface, surface of concrete, etc.) is a better reference point. In any GPR system, the signals are generated within the instrument and processing itself occurs within the instrument. The signals must travel to and from the antennas and may be affected by both propagation times and delays in the various blocks within the instrument. These delays affect the raw data, particularly with respect to target location information. In ground-coupled radars, the delays are entirely due to propagation within the components of the instrument. If the assessment is performed using an air-coupled antenna, the distance to ground relative to wavelength is considerable and results in additional delays. In addition, the environmental conditions under which the test is performed, such as differences in temperature, can influence the system, the speed of acquisition, and the speed of propagation. For accurate measurements, it is necessary to place reflections in their true space locations, that is, to identify the true location of targets. Due to all of these effects and delays, the uniform solid layer in the radargram may not be properly positioned in time (and hence in space). Because it does not contribute to the assessment, it can be removed by "repositioning" it to its true reference, which is the surface being scanned. A common way to correct the signal is to analyze zero crossings to identify the first inflection or the first positive or negative peak. Time zero correction is the procedure to bring the signal to the real origin in order to improve the detection of targets in their proper depth by matching the surface position. Raw data with different time zero values among *A*-scans can indicate problems with the system. Figure 4.15 shows an example of an *A*-scan. In this example, the first reflection occurs after time zero. It is important to shift all the *A*-scans to the same time zero in order to better estimate depths of targets. This is done by interpolation on all traces of an assessment. Figure 4.16 shows the two-way travel time of the reflected

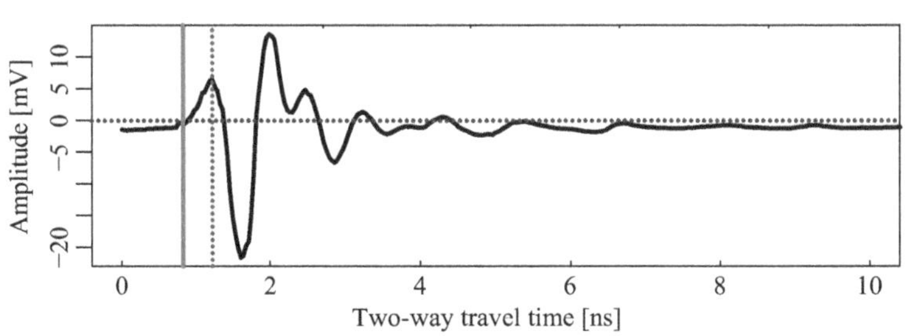

Figure 4.15 An A-scan from measurement. The vertical solid line shows the time zero location

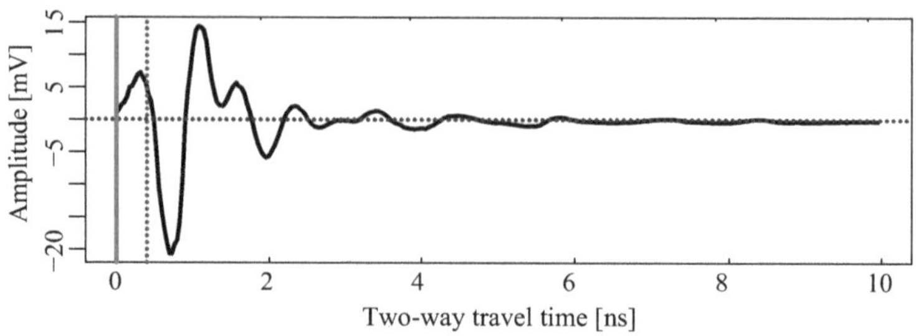

Figure 4.16 Time zero corrected A-scan

field from the host medium after this interpolation was applied on the data in Figure 4.15. This data can be used to approximate depth measurements by using the velocity of the wave (assuming of course that it is known or can be estimated).

4.7.4.3 Background subtraction

Another simple preprocessing operation that can have significant impact on the detection of targets is background subtraction. The transmitted signal undergoes multiple interferences and is affected by anything in its path from the transmitting to the receiving antenna, effects that are unrelated to the target itself. The main issue involved in detecting different types of targets is the discrimination from interference effects from the environment in which the target is embedded. Also, the signals due to background effects are usually more intense than the signals from targets. In many practical situations, the background signals can obscure the signals from targets. Background signals are therefore a part of the assessment that is not useful in the detection of buried objects. Background subtraction can improve detection and discrimination between scattered targets. In addition, it can minimize the problem of false positives.

There are a number of ways by which background subtraction can be affected, depending on the goals to be achieved. The goal may be removal of random noise, electronics noise, or simply the removal of an abrupt background. Since the raw data are composed of many *A*-scans, one has to weigh if applying background removal to the whole data is the better option. For instance, if the objective is to detect different types of buried pipes (i.e., metal and plastic) in one test, background removal tends to aggregate intense reflections from one type of pipes (metal) with others (those from plastic pipes) distorting the assessment.

Background subtraction can be done by averaging and subtraction in each *A*-scan. This can be done because the causes of the interference are deterministic. It can also be done on the multiple *A*-scans that make up the *B*-scan or on the complete underground image.

Heterogeneous changes in the medium (when these changes exhibit high contrast) can obscure reflections from the space immediately below the interface between air and the material under test. It is possible to subtract these reflections by calculating the time of arrival of the waves. In this case, the operator has to identify the clutter from the *A*-scans in the *B*-scan and apply the correction without the loss of information from

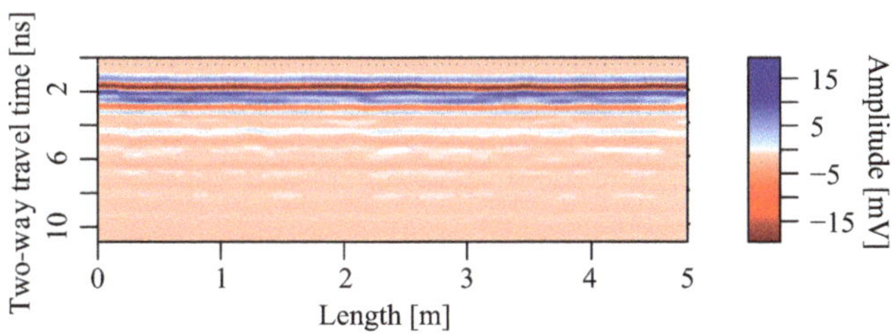

Figure 4.17 Background removal

early reflections from the subsurface. Figure 4.17 shows a *B*-scan with background subtraction where horizontal improvement of the image can be seen. In this case, the noise may have originated from the system, surface reflections, or EMC issues.

4.7.4.4 Downsampling

Although background subtraction is a simple and effective way to reduce clutter in the GPR raw data, it fails to completely remove clutter in situations in which the interference is time varying [26]. Downsampling is a process of reducing the sampling rate of a sequence (i.e., an *A*-scan sequence) by a factor M [Equation (4.14)]. The method assumes, implicitly, that the original data must have been sampled beyond the Nyquist rate, or loss of data may occur due to the downsampling process. Downsampling expands the content of the signal in the frequency-domain. As a consequence, the choice of the downsampling factor must be such that it avoids aliasing in the downsampled sequence. Some applications may require a pre-filtering step before downsampling. Downsampling is also effective in reducing the number of unknowns for interpretation algorithms or pattern recognition as it has the net effect of reducing the sampling rate.

$$x[n] = x[nM] \tag{4.14}$$

4.7.5 Basic signal processing

Basic signal processing consists of those techniques applied to data that require operations beyond those afforded by the preprocessing steps. Their use also requires some knowledge and interaction on the part of the operator. These techniques are typically needed because, after the preprocessing step, the raw data still contain noise that can obscure targets or generate false alarms. Some a priori information about the assessment is typically needed, for instance a gain parameter, the transmitted signal, or the cutoff frequency of a filter. This information is supplied to the system by the operator. Figure 4.18 illustrates a GPR survey that was used to collect data to be used in the following subsections to demonstrate the concepts discussed. Figure 4.19 shows the raw radargram obtained in this survey.

In a GPR assessment such as that shown in Figure 4.18, part of the power transmitted into the ground couples directly from the transmitter to the receiver. This is also called the direct air wave. In addition to the reflection from underground, the

Figure 4.18 Radar survey on the ground. The grid is used to keep scans straight and parallel

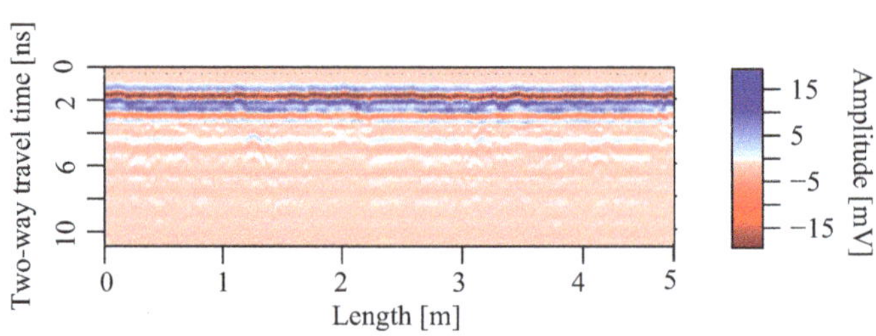

Figure 4.19 Radargram from the GPR assessment in Figure 4.18

power scattered from the ground (called the ground wave) will be added. The direct waves and the ground waves do not contribute to the detection of underground targets and can be removed to improve detection. Most commercial GPR systems include filters that can be used after a survey is complete. If the goal is to detect underground layers or horizontal changes or contrasts, these techniques are not the best solution. Figure 4.18 is an example of how complex a GPR assessment is. Even in this case, where the surface is clean and prepared (Figure 4.18), the raw data show considerable variations. In practice, the soil is not flat, it contains inhomogeneities such as grass, rocks, roots, and other materials that interfere with the probing EM wave, all of which have an impact on the raw data.

Figure 4.19 is the raw signal as recorded by the receiving antenna and includes the direct air wave, ground wave, and all noise in the environment, including noise from RF transmitters (radio stations, mobile towers), power systems, and the like. If, for example, a strong RF transmitter operates close to the GPR assessment, the signals from this source may be stronger than reflections from underground and can corrupt the data. Without treatment, it is more difficult or sometimes impossible to find underground targets because the power in early-time signals is always higher than the late-time signals from deeper sources. To handle some of these signals, there are a number of techniques that can be used and these will be discussed in the following sections. Some are basic, whereas others are more advanced. These, if properly used, can increase the likelihood of detecting underground targets considerably.

4.7.5.1 DC shift removal

One of the effects of air waves and ground waves is a shift in the DC level of *A*-scans (and as a consequence of *B*- and *C*-scans). In some applications, the DC shift may be small but it is not negligible. In other applications, the shift may be significant. To better analyze *A*-scans, a zero mean in the signal is desirable. Some GPR manufacturers provide software to correct the received signal for DC shift. DC shift removal can be considered as a one-dimensional filter. As an example, (4.15) applies an average-subtraction routine capable of calculating the DC level of the signal.

$$x_{dc}[n] = x[n] - \frac{1}{N}\sum_{k=1}^{N} x[k] \tag{4.15}$$

where $x_{dc}[n]$ is the DC level to be removed from the signal, $x[n]$ is the field data, and N is the number of samples in the trace.

Figure 4.20 illustrates an *A*-scan taken from Figure 4.14 (i.e., this is one trace in the *B*-scan in the figure). The dashed line is the DC shift calculated from the *A*-scan. Figure 4.21 shows the *B*-scan after DC shift removal. In general, DC shift removal is used to enhance dewow filters (to be discussed in the next subsection).

4.7.5.2 Dewow filtering

Dewow filtering is the process of removing DC offset drift or low-frequency content generated by the GPR system, including electrostatic or induced fields in the GPR and/or due to instrumentation dynamic range limitations. The difference between Dewow filtering and DC shift removal stems from the fact that the shift in signal in this

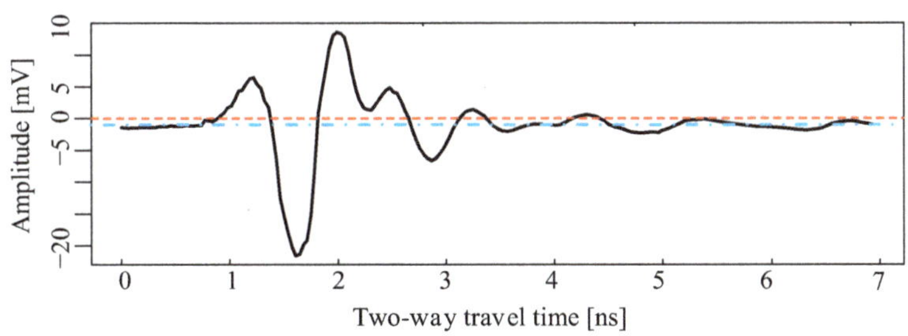

Figure 4.20 An A-scan trace taken from the survey in Figure 4.14

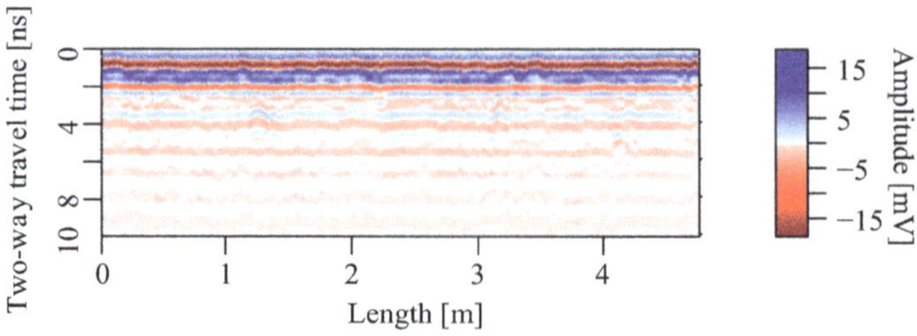

Figure 4.21 The B-scan in the figure after DC shift removal

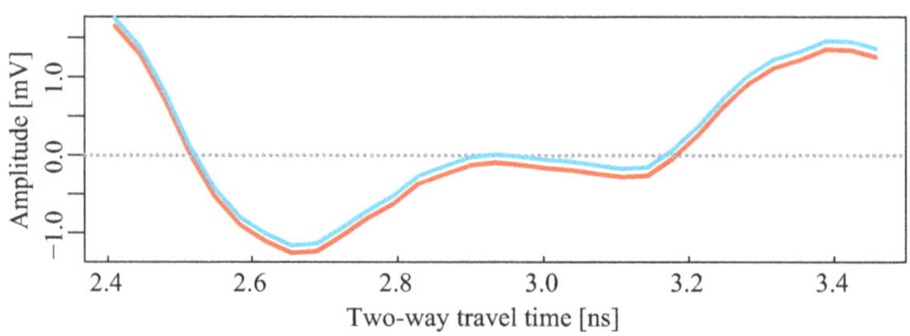

Figure 4.22 Dewow filtering before dewowing (red line) and after dewowing (blue line)

case is not constant. This phenomenon can obscure actual reflections or diffraction from targets. Removing low-frequency components can be achieved using several types of filters such as median and mean filters or Gaussian filters. Figure 4.22 shows the *A*-scan before (dashed line) and after dewowing.

Figure 4.22 illustrates only the part of the signal affected by the wow which may be caused by signal saturation, instrumentation limitations, or inductive effects. Since only part of the signal is modified, it is recommended to apply dewow filters before other procedures such as gain or attenuation are undertaken.

4.7.5.3 Signal amplification and attenuation

In practical use, the signals coupled into the domain of interest encounter media that vary from very low to high loss. It may also be possible in some cases to assume lossless media. However, in almost all cases, an exponential attenuation of fields and power takes place (lossy media) as was defined in Chapter 2 through the concept of attenuation constant, α (measured in Nepers/m or in dB/m). In addition, scattering in directions in the medium that cannot be received, and spread of the signal all contribute to the basic fact that the received signal, even in the absence of noise is typically very low. This is particularly so for signals from deeper targets. To compensate for this, it often becomes necessary to amplify weak signals and to attenuate strong signals.

The signals reflected from underground targets generate well-defined patterns in the resulting A-scans and, as a consequence, in B-scans. Typically, the patterns are hyperbolic or straight segments. This effect can be seen in Figure 4.23, which shows the image due to 12 cylindrical targets of different diameters and material properties buried at various depths underground (the result shown here was obtained by simulation and will be discussed in detail in Chapter 5). The exact shape and contrast seen in the image depends on the target. For example, a metal target will produce a clear, intense hyperbola, whereas a fiber optic cable or an air gap in concrete will produce a fainter, less pronounced hyperbola. Similar statements apply to dimensions of targets and the depths where they are located.

Straightforward amplification is an obvious choice but, given these patterns in the reflected signals and because the loss over the path is not constant, a proper method of enhancement of the signal should be selective with higher amplification for signals from deeper sources (longer time delays at the receiver). To take these considerations into account, one may apply a time-varying enhancement to the data so that some characteristics or targets stand out from the rest of the signal. This is not a simple procedure since the user cannot predict exactly the behavior of materials underground. For these reasons, there are different formulas that have been proposed to compensate the signal under different conditions. Two examples are given in (4.16) and (4.17), representing power gain and exponential gain, respectively. The power parameter, p, in the functions is selected such as to address the attenuation experienced by the

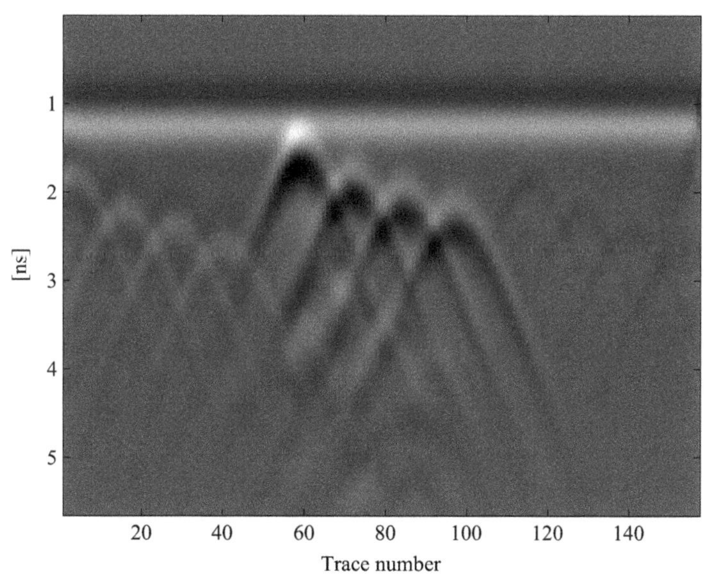

Figure 4.23 Simulated GPR scan with 12 targets. Note the hyperbolic pattern produced by each target

signal and compensate for the attenuation. In other cases, the power gain enhancement through (4.16) may produce better results.

$$x_g(t) = x(t) \cdot t^p \tag{4.16}$$

$$x_g(t) = x(t) \cdot e^{pt} \tag{4.17}$$

Figure 4.24 shows the effect of an exponential gain function applied to the GPR assessment performed in Figure 4.18 showing the enhancement in amplitude with time (and hence with depth). Note as well the patterns in the figure. An image produced after applying this enhancement will emphasize deeper effects and thus improve detection.

There are other types of gain functions that can be applied. These include automatic gain control (AGC) and its derivatives (i.e., Gaussian-tapered AGC), inverse power decay, and spherical and exponential compensation. It is also possible for the user to customize gain functions to address specific conditions, usually based on some knowledge acquired during tests.

The choice of what method to use must be made on the basis of some prior knowledge of the structure under test. In addition, the user must be careful when noise is present, since it can be enhanced along with the signal. In some regions of the raw data, the user may want or need to attenuate instead of amplifying the signal. For instance, if different types of targets are present at the same depth and in the same host medium, the pattern's shape and contrast will differ. Proper gain levels (amplification or attenuation) may help in distinguishing between different types of targets such as long or linear targets, cylindrical pipes, cables, rebar, tree roots, and so on.

Attenuation correction is widely used in impulsive GPRs. However, SFCW GPR systems require different gain functions [27], since gain functions used for impulsive radar are not effective. This is because in SFCW radar the signal is analyzed using time, waveform, and phase information extracted from the samples through a coherent mechanism, in addition to amplitude information.

Selection of an appropriate gain function is a difficult task since it depends on the type of assessment and the operator skill in addition to the conditions underground, the type of target, noise, type of radar, and skill of the operator. Some experimentation can also be helpful in settling on a proper gain function.

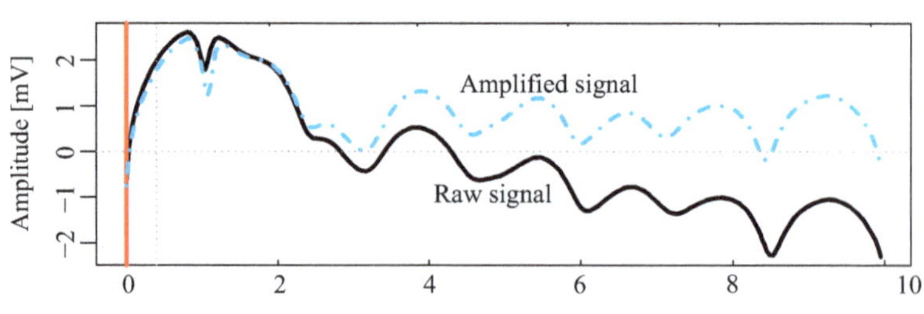

Figure 4.24 Effect of exponential gain function (filter) on a signal

4.7.5.4 Deconvolution

Figure 4.25 illustrates a generic GPR system abstraction. The unprocessed GPR data can only provide a distorted image at best. This is due to the fact that the final signal is a convolution of the waveform generated (using amplitude or frequency modulation) and the impulse responses of the various blocks in the system, including media and antennas.

The immediate approach that imposes itself is deconvolution. Deconvolution is an inverse procedure applied to the received signal in order to isolate the impulse response of the target under investigation from other responses in the waveform. It can be considered a temporal filter used to reverse the convolution effects [28], i.e., it is used to remove the source waveform effect from the raw data. Its purpose in the current context is to improve the vertical resolution by filtering unimportant features [29]. The received signal $x[t]$ is the convolution between the transmitted wave $s_{tx}[t]$ and the target response in host medium $s_{target}[t]$:

$$x[t] = s_{tx}[t] * s_{target}[t] \tag{4.18}$$

There are a number of situations where deconvolution filters can be applied. In some scenarios, the GPR energy can be confined to specific areas. If the scenario consists of layered media with layers of high reflectivity (e.g., ice over water), the signal may reflect back and forth several times. These reflections are often referred to as ringing. The multiple reflections are detected by the receiving antenna with delays imposed by the thickness of layers causing the user to believe that they are coming from deeper regions. Deconvolution filters compress the data in order to find the original reflection and eliminate its multiple echoes.

The convolution operation in time is equivalent to multiplication or division in frequency. For this reason, it is much simpler to apply deconvolution filters in the frequency-domain using spectral division deconvolution. However, it is important to take some precautions when using this filter. It is possible that important data from the assessment can also be removed. For instance, if there are reflecting layers such as metal below the layer causing the ringing, a deconvolution filter may remove those reflections from the data. The use of convolution methods has proven particularly effective when the user has prior information about the subsurface. For this reason,

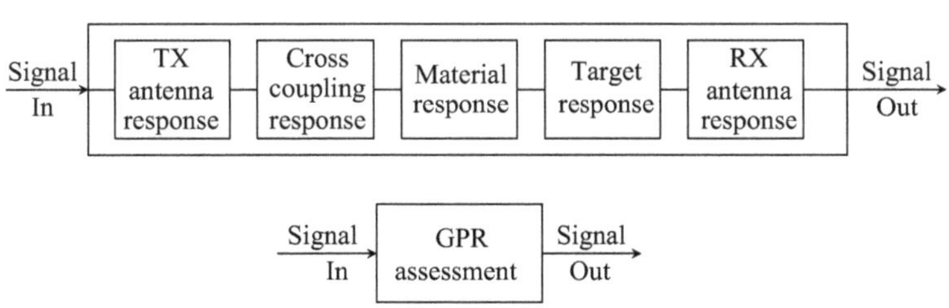

Figure 4.25 Functional blocks in a GPR assessment

deconvolution has been largely employed in seismic applications. In other situations, the mixed-phase GPR wave becomes more difficult to analyze.

4.7.5.5 Matched filters

In any use of GPR, one has to contend with low amplitudes and data corrupted by clutter, noise, and interferences. Weak scattering from targets, lossy media, and spread of signals all contribute to this basic difficulty in radar assessment. Even for lossless homogeneous materials, the received signal is often too weak to analyze.

A powerful operation, borrowed from the wider radar community applied to GPR, is the matched filter. The main goal of this filter is to compare the received signal (with all its components, including clutter, interference, and noise) and the transmitted signal and generate an output that is maximized when a reflection from a possible target appears. The structure of the matched filter is shown in (4.19).

$$x_{MF}[n] = \sum_{k=-\infty}^{\infty} x[n]\, h[n-k] \tag{4.19}$$

This operation is closely related to convolution and correlation. However, there is a difference. The matched filter computes true correlation between the received and transmitted signal for the duration of the whole assessment. The filter can identify signal components that are highly correlated with a reversed time shifted version of the transmitted signal even in the presence of interference and noise. Using this type of filter, it is possible to identify the target signals in the received data by computing the travel times of the highly correlated responses.

4.7.5.6 Band-pass filtering

The receiving antenna and the receiver circuitry in most GPRs are designed to collect signals in a wide frequency range. In high-resolution impulsive GPR systems, in particular, the frequency bandwidth is particularly large by design since GPR resolution is related to the bandwidth used in the assessment. Some radars operate as UWB systems (UWB is defined by the IEEE Standard 686 as bandwidth over 25% of the center frequency or larger than 500 MHz [30]). Because of that, and depending on the application, the system may be subjected to interference from a variety of devices working in the same frequency range such as mobile phones, radio and TV stations, Wi-Fi routers, and EM signals from many other sources. The signals from these sources, some of which may be quite intense, can degrade the SNR in the assessment. Reduction or elimination of noise is an important and difficult objective when using GPR. To achieve an optimal SNR, it is necessary to understand the differences between signal and noise in order to apply a methodology to eliminate (reduce) noise without the loss of information that is important for a successful interpretation.

The use of very short pulses imposes some particular challenges to GPR system developers. Pulse-shape distortion, reduction in multipath propagation, high-frequency synchronization, the avoidance of RFI, and low transmission power are some of the issues that make UWB applications more complex and more challenging. In addition, with more demand for high-resolution GPR applications, there is a growing need for frequency allocation for UWB radars. Since the EM spectrum is crowded and growing more crowded with demands for spectrum from other applications, the allocation for new systems is an important consideration due to possible (or even likely) RFI [31].

Since time-limited signals generate a band-unlimited spectrum, some devices can insert signal components in the raw data at frequencies outside the GPR operating range. Band-pass filters are applied to improve the SNR by removing noise components incorporated into the raw data from external devices. Noise components outside a specific frequency range can be eliminated through use of band-pass filters. In addition, the signal amplitude can be recovered in the band-pass range restoring power lost due to attenuation.

However, the choice of the cutoff frequencies for the band-pass filter should be considered carefully and must take into account the application and its goals since important information can be lost if they are in, or partially overlap, the components filtered. After analyzing the signal spectra and considering the antenna center frequency, a bandwidth for the filter must be defined. A bandwidth of 1.5 times the center frequency is routinely used in practice [8] but specific applications may require a narrower or wider band. Figure 4.26 shows a band-pass filter applied to a trace (an *A*-scan) from the data and Figure 4.21 together with the unfiltered data. In this example, a band-pass filter with cutoff frequencies of 400 MHz and 2.8 GHz was applied. Both of these frequencies are outside the GPR system bandwidth.

Figure 4.27 illustrates how the filtered spectrum can influence the interpretation of the data. Comparing Figure 4.21 with Figure 4.27, it is possible to see the effect of removing the high- and low-frequency noise.

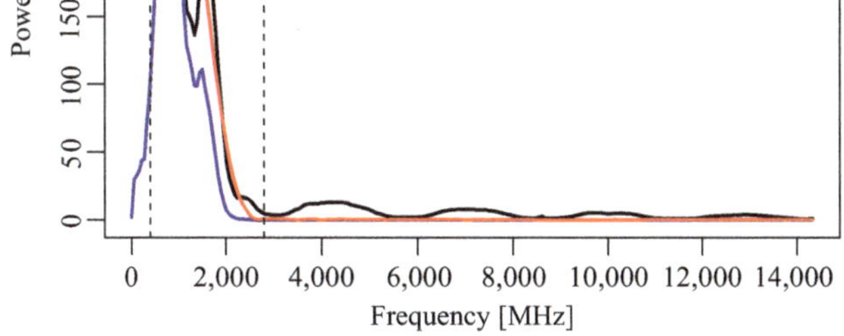

Figure 4.26 Band-pass filtering. Cutoff frequencies are 400 MHz and 2.8 GHz

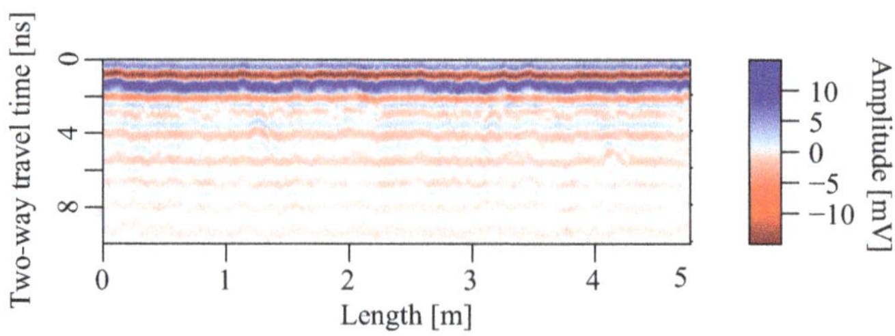

Figure 4.27 B-scan after band-pass filtering with the filter in Figure 4.26

4.7.5.7 Phase velocity analysis

Phase velocity spectra are used in GPR assessments to extract velocity information from fields scattered from targets. From velocity information, one can extract dielectric constants in order to characterize the medium. This can be done using the assumption that the medium has negligible conductivity, and the permeability is that of vacuum. The idea is obvious if the medium is uniform but the task becomes more difficult when the subsurface consists of layers with different properties. In this situation, the phase velocity in each layer is different and the two-way travel time of the target signal more difficult to estimate.

The goal of velocity analysis or time-to-depth conversion is to improve SNR. It is a well-known procedure particularly common in seismic work where material data acquisition mode plays an important role. Phase velocity is most often measured using the CMP mode of data acquisition as discussed in Section 4.6.3, since it works even in the presence of diffracting or reflecting targets. In CMP, the transmitter and receiver change the offset distance between one another during the assessment remaining symmetrical about the target (the midpoint). In CMP measurements, assuming that the medium is lossless, the two-way travel time can be estimated as a hyperbolic function of the antenna offset as described in the following equation:

$$time\,(x) = \sqrt{\left(\frac{x}{v_p}\right)^2 + \left(\frac{2d}{v_p}\right)^2} \tag{4.20}$$

where v_p is the electromagnetic wave velocity (phase velocity) in air, d is the depth of the horizontal medium, and x is the antenna offset in the transmitter and receiver. This is a robust method for subsurface spatial control. However, the accuracy and precision of depth estimates using velocity analysis are dependent on the knowledge of the subsurface phase velocity. Precision of the travel-time observations is also affected by the configuration of the GPR system, including frequency, antennas configuration, and so on. As with any method, prior knowledge of subsurface properties, even if partial, can simplify analysis and lead to better velocity estimates and hence to the location and evaluation of targets.

CMP measurements are often used to provide data for velocity analysis. But there are other empirical ways to determine the phase velocity, some more accurate than others. An obvious method is the use of phase velocity tables for known types of media such as specific soils or concrete. Another is based on verifying the EM properties of the materials as discussed in Chapter 2 and determining their phase velocities compared with the host medium. Since it is difficult to characterize the substrate for every measurement performed, this method can lead to errors in estimating depth.

Another empirical possibility is a kind of destructive testing: implant scatterer. The user buries an object in the host medium and collects data using the GPR to verify the two-way travel time and then calculates the velocity. Since the user modifies the subsurface (at least locally), this method can only be applied to a small number of applications and may be entirely inappropriate in others (such as in archaeology). In addition, the host medium may not be entirely homogeneous or may vary from location to location or other depths leading to lateral and vertical inaccuracies [32]. Nevertheless, under some conditions and in particular if these conditions are common (such as sands, clays), this is a simple and often reliable method of collecting new data or verifying data from other sources.

4.7.5.8 Migration

Migration is a focusing technique applied in GPR for lateral enhancement. It attempts to bring EM waves into focus by exploiting measured data to map the subsurface. The consequences of migration can be summarized by the following properties [33]:

1. improvement of lateral resolution;
2. correction for the distorted position of reflectors; and
3. correction for the distorted amplitude of reflections.

A number of migration techniques that were initially developed for seismic work were adopted and adapted to GPR. The choice of a technique has to take into account parameters such as the target characteristics, the available measurement data, and complexity of the assessment. Migration involves complex mathematical relations and as GPR has become more mainstream, coding computationally efficient migration algorithms has become more frequent.

An interesting example is the reverse-time migration technique [34], designed to find the exact location of targets in heterogeneous media. The development of this algorithm is based on the notion of a matched filter, a concept used extensively in radar applications. The matched-filter concept can be explained as a correlation of the received signal with the expected or estimated signal from a specific target as was previously explained in Section 4.7.5.5. If this correlation produces a large value, then it is likely that the target is present at the location indicated. Using this algorithm, an image can be viewed as a back-propagated wave-field reconstruction of the dielectric contrast within the host medium [34]. The final migrated data, $S(\vec{r})$, for a bistatic radar configuration can be obtained from the following relation:

$$S(\vec{r}) = \sum_{n=1}^{N} \sum_{m=1}^{M} E_{mn,bp}(\vec{r}) * E_{inc}(\vec{r})\,|_{t=0} \tag{4.21}$$

where the subscripts m and n denote the field due to the mth transmitter and nth receiver. Therefore, this expression is interpreted as the intersection (convolution) of the back-propagated field ($E_{mn,bp}$) with the incident field (E_{inc}). This technique will be further discussed in Chapter 6 in the context of imaging.

4.7.5.9 *F–k* wave filtering

F–k was filtering is a method also adapted from seismic applications whereby the goal is to separate a desired wave type from the readings (in seismic applications these are known as up- and down-propagating *P* (pressure) and *S* (shear) waves). For GPR applications, one possible objective can be to separate the scattered field from targets from the incident wave [35]. A comprehensive representation is obtained by filtering the data in the frequency-domain with a given filter function. Thus, it is necessary to map the time-space domain into the frequency-wavenumber domain (hence the name *F–k*). This can be done by applying the Fourier transform to *B*-scans. The filtered result $X_f(k, \omega)$ is the product of a filter function $H(k, \omega)$ and the 2D Fourier transform of the radargram $x(x, t)$ or $X(k, \omega)$:

$$X_f(k, \omega) = H(k, \omega) \cdot X(k, \omega) \tag{4.22}$$

The function $H(k, \omega)$ is a filter function, the purpose of which is to reject information from data such as the incident field or noise. However, like all filter functions, it is only possible to reduce or attenuate rather than completely eliminate all incident waves. As a result, some undesired information is left in the resulting image. For best results, the method relies on the experience of the operator to determine the spatial band of the filtering function. Background removal can also be considered a spatial filtering technique and, therefore, *F–k* filtering is often used to remove clutter and noise from GPR readings. Filtering can be applied to all traces or, selectively, to regions of interest.

Figure 4.28 shows the GPR assessment of a gravel-covered surface, and Figure 4.29 shows the raw data collected over that surface in Figure 4.28. These data are used as an example for the application of *F–k* filtering.

Figure 4.28 GPR scenario with rocks at the interface

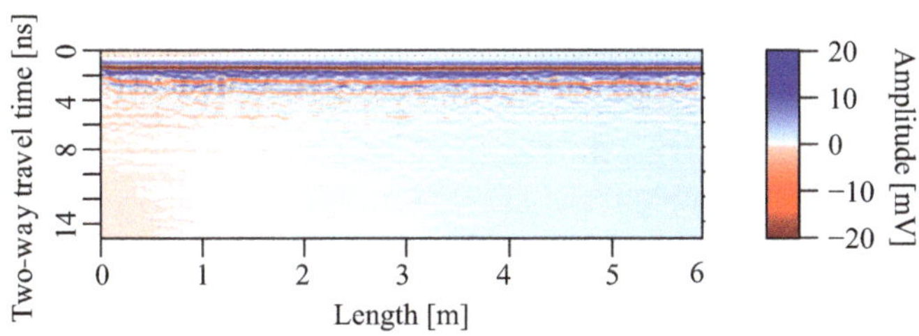

Figure 4.29 B-scan of the GPR assessment in Figure 4.28

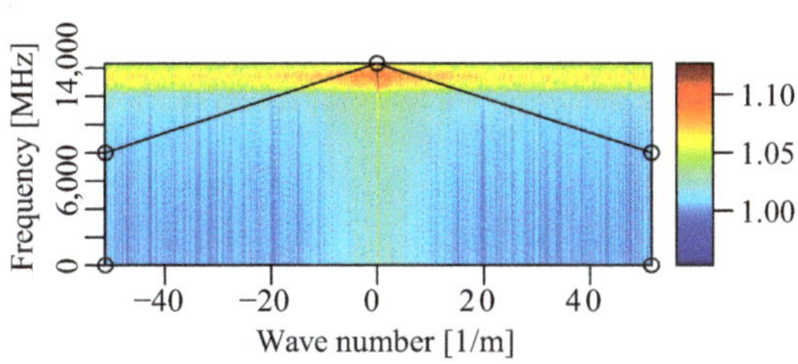

Figure 4.30 F–k plot of the data in Figure 4.29

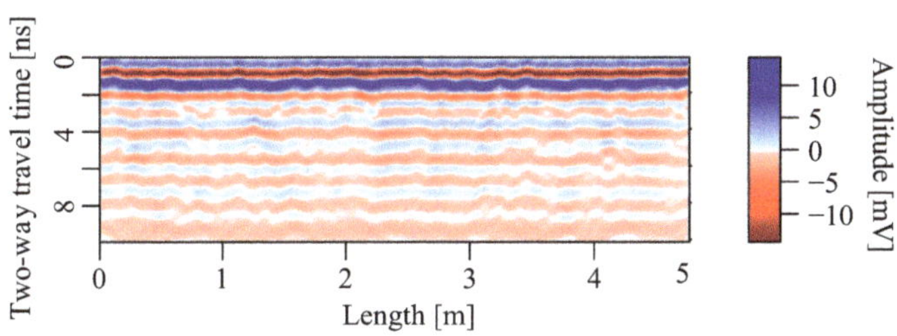

Figure 4.31 Filtered radargram showing higher contrast and the strong signals due to the gravel layer

Figure 4.30 is a plot in the frequency-wavenumber (F–k), plane, with the amplitude of the GPR data shown by the color scale on the right of the plot. In Figure 4.31, it is possible to see the strong interface reflection from the gravel layer that was not evident in the raw data in Figure 4.29.

4.7.6 Advanced signal processing

As stated before, different applications may require different signal processing procedures. The applications themselves may be as simple as detection of a pipeline for maintenance purposes or as complex and critical as detection and neutralization of landmines. The detection of buried mines, especially antipersonnel mines, is one of the most difficult problems GPR is tasked with. Part of the difficulty is the fact that landmines are developed to mimic the ground characteristics. In addition, they may be buried in various terrain and soil conditions making prediction of host material characteristics impractical. The main goal is to distinguish nonmetal targets from the ground with a very low false alarm rate and a high detection rate in real time. The signal processing methods described earlier increase computation significantly due to the complexity of the problem and cannot typically yield data in real time. Fortunately, there is a class of spectral estimation techniques that can deal with complex applications in real time [36]. These more advanced signal processing techniques can be summarized as follows:

1. maximum likelihood method;
2. minimum entropy method;
3. multiple signal classification;
4. estimation of signal parameter via rotational invariance technique;
5. principal component analysis (PCA);
6. Kalman filtering;
7. wavelet packet decomposition; and
8. Monte Carlo methods.

The PCA method will be discussed in Chapter 6, but it is beyond the scope of the present work to dwell into the details of these more advanced methods. Part of the reason for excluding them is their complexity that requires significant mathematical tools but mostly because the purpose here is primarily to look at general-purpose methods that the user might encounter.

4.8 Summary

The present chapter dealt with a number of topics starting with the operation and classification of ground-operating radar followed by system requirements for a successful implementation of a GPR. These not only include hardware components of the transmitter and receiver, including waveforms, that must be addressed but also operational requirements such as scanning modes. The main body of the chapter, however, deals with all important aspects of signal processing. The importance of signal processing cannot be overstated: it is through signal processing that significant improvements in detection and characterization of targets can be made, once the system is assembled. This will be followed by image processing methods in Chapter 6 but before that a discussion on numerical methods as auxiliary tools is introduced in the following chapter.

References

[1] Skolnik MI. Radar Handbook, 3rd ed. Electronics Electrical Engineering. McGraw-Hill Education, New York, NY; 2008.

[2] Electronic Code of Federal Regulations, Section 15.509 – Technical requirements for ground penetrating radars and wall imaging systems. Accessed: 2021-02-21. https://nam03.safelinks.protection.outlook.com.

[3] Electronic Code of Federal Regulations, Section 15.521 – Technical requirements applicable to all UWB devices. Accessed: 2021-02-21. https://nam03.safelinks.protection.outlook.com.

[4] Balanis CA. Antenna Theory – Analysis and Design. John Wiley & Sons, Hoboken, NJ; 2016.

[5] Reeves B. Noise modulated GPR: Second generation technology. In: Proceedings of the 15th International Conference on Ground Penetrating Radar; 2014:708–713.

[6] Raimundo X, Salous S and Floranelli F. Frequency modulated interrupted continuous wave signals in different radar imaging applications. In: 2014 XXXIth URSI General Assembly and Scientific Symposium (URSI GASS); 2014: 1–4.

[7] Doerry A. Noise and Noise Figure for Radar Receivers. DOE Technical Report; 2016. Accessed: 2021-02-22. 2021.https://doi.org/10.2172/1562649.

[8] Jol HM. Ground Penetrating Radar Theory and Applications. Elsevier Science, Amsterdam; 2008.

[9] Fowle EN. The Design of Radar Signals. Technical Report No. ESD-TR-65-97. United States Air Force; 1965.

[10] Zhang L, Wei N, and Du X. Waveform design for improved detection of extended targets in sea clutter. Sensors, 19 (3957);2019:1–10.

[11] Garren SA, Osborn MK, Odom AC, Goldstein JS, Pillai SU and Guerci JR. Enhanced target detection and identification via optimised radar transmission pulse shape. IEE Proceedings – Radar, Sonar and Navigation, 148;2001: 130–138.

[12] Romero R and Goodman N. Waveform design in signal-dependent interference and application to target recognition with multiple transmissions. IET Radar, Sonar & Navigation, 3;2009:328–340.

[13] Aubry A, Carotenuto V, De Maio A, Farina A, and Pallotta L. Optimization theory-based radar waveform design for spectrally dense environments. IEEE Aerospace and Electronic Systems Magazine, 31;2016:14–25.

[14] Uduwawala D. Gaussian vs differentiated Gaussian as the input pulse for ground penetrating radar applications. In: 2007 International Conference on Industrial and Information Systems; 2007:9–11. DOI: 10.1109/ICIINFS.2007.4579173.

[15] Voudras A. Gaussian bases for radar signal analysis. Signal Processing, 20 (2);1990:163–169.

[16] Ricker N. The form and laws of propagation of seismic wavelets. Geophysics, 18 (1);1952:10–40.

[17] Wang Y. The Ricker wavelet and the Lambert W function. Geophysical Journal International, 200 (1);2015:111–115.

[18] Noon DA. Stepped-Frequency Radar Design and Signal Processing Enhances Ground Penetrating Radar Performance. The University of Queensland; 1996.

[19] Eustice D, Baylis C, Cohen L, and Marks RJ. Effects of power amplifier nonlinearities on the radar ambiguity function. In: 2015 IEEE Radar Conference (RadarCon); 2015:1725–1729.

[20] Wiggins JW. Kirchhoff integral extrapolation and migration of nonplanar data. Geophysics, 49 (8);1984:1239–1248.
[21] Lehmann F and Green AG. Topographic migration of georadar data: Implications for acquisition and processing. Geophysics, 65 (3);2000:836–848.
[22] Berkhout AJ. Wave field extrapolation techniques in seismic migration, a tutorial. Geophysics, 46 (12);1981:1638–1656.
[23] Moran ML, Greenfield J, Arcone SA, and Delaney AJ. Multidimensional GPR array processing using Kirchhoff migration. Journal of Applied Geophysics, 43 (2);2000:281–295.
[24] Smitha N, Ullas Bharadwaj DR, Abilash S, Sridhara SN, and Vipula Singh V. Kirchhoff and F-K migration to focus ground penetrating radar images. International Journal of Geo-Engineering, 7 (1);2016. DOI: 10.1186/s40703-016-0019-6.
[25] Annan AP. Chapter 11: Ground-Penetrating Radar. Society of Exploration Geophysicists, Denver CO; 2005; p. 357–438.
[26] Gupta M, Vuksanovic B, and Hidzir H. Discarding unwanted features from GPR data using downsample-upsample method. In: 2013 14th International Radar Symposium (IRS). vol. 2; 2013:829–834.
[27] Liu T, Zhu Y, and Su Y. Method for compensating signal attenuation using stepped-frequency ground penetrating radar. Sensors, 4 (18);2018:1366.
[28] Benedetto F and Tosti F. A signal processing methodology for assessing the performance of ASTM standard test methods for GPR SYSTEMS. Signal Processing, 6;2016.
[29] Ciampoli LB, Tosti F, Economou N, and Benedetto F. Signal processing of GPR data for road surveys. Geosciences, 2 (9);2019:1–20.
[30] IEEE Standard for Radar Definitions. IEEE Std 686-2017 (Revision of IEEE Std 686-2008); 2017. p. 1–54.
[31] Davis ME. Frequency allocation challenges for ultra-wideband radars. IEEE Aerospace and Electronic Systems Magazine, 28 (7);2013:12–18.
[32] Jacob RW and Urban TM. Ground-penetrating radar velocity determination and precision estimates using common-midpoint (CMP) collection with hand-picking, semblance analysis and cross-correlation analysis: A case study and tutorial for archaeologists. Archaeometry, 58 (6);2016:987–1002.
[33] Berkhout AJ. Seismic Migration: Imaging of Acoustic Energy by Wave Field Extrapolation. Vol. 14, pt. 1 in Developments in Solid Earth Geophysics. Elsevier Scientific Pub. Co., Amsterdam; 1982.
[34] Leuschen C and Plumb R. A matched-filter-based reverse-time migration algorithm for ground-penetrating radar data. IEEE Transactions on Geoscience and Remote Sensing, 6 (39);2001:929–936.
[35] Hayashi N and Sato M. *F–k* filter designs to suppress direct waves for bistatic ground penetrating radar. IEEE Transactions on Geoscience and Remote Sensing, 48 (3);2010:1433–1444.
[36] Ikuo A and Shrestha S. Signal processing of ground penetrating radar using spectral estimation techniques to estimate the position of buried targets. EURASIP Journal on Advances in Signal Processing, 2003;2003.

Chapter 6
Pattern recognition

6.1 Introduction

There is an indefinite number of different scenarios when it comes to nondestructive testing (NDT) using ground penetrating radar (GPR). Each scenario has its own properties with respect to system specification, material properties, and environment conditions. In the preceding chapters, it was made amply clear that electromagnetic (EM) wave propagation results in complex responses even from simple configurations of materials and targets in any given scenario. In GPR, the responses from the testing environment are used to identify and quantify targets and other features that may be of interest. The raw data must be processed in a way that provides useful, dependable information that can be acted upon. Pattern recognition is a very important step in the overall purpose of extracting useful information from the received signals.

In pattern recognition of GPR data, one is given a training set (received by the receiving antenna and recorded for further processing) and attempt to predict the sources that created the reflections (targets). This is considered as an EM inverse problem, the aim of which is to determine a finite number of parameters needed to characterize targets in a given medium by identifying the electric and geometric properties of the targets. In general, this is a very difficult problem because the information available from reflections is almost never sufficient for estimation of properties, giving rise to the need for some a priori information about the configuration and its constituents. On the other hand, a large number of data that ostensibly can fill in the gaps in information can bring the problem to instability.

The purpose of pattern recognition techniques is to improve the GPR assessment and diminish the importance of human interpretation by solving the EM inverse problem associated with GPR tests. In general, pattern recognition is applied after signal processing techniques have been brought to bear on the data for the purpose of cleaning, enhancing, and improving the signal. To be effective, these techniques should be cost effective, adaptable to different applications, provide a reasonably fast response for large amounts of data, and provide a desirable false alarm response or probability of detection.

Given the complexity of GPR systems and data available, it must be admitted at the outset that it is virtually impossible to develop a technique that can classify and distinguish buried targets in GPR assessments of all possible NDT configurations.

It is therefore common to adopt various strategies for various applications or classes of applications.

Each pattern recognition step may use a number of different techniques that include, but are not limited to, principal component analysis (PCA), discriminant analysis, Markov models, decision trees (DTs), *k*-nearest neighbors (*k*-NNs), edge histogram descriptors, spectral features, Bayesian classifiers, geometric features, texture features, neural networks (NNs), and machine learning. The problem of obtaining information on buried inclusions such as shape and material properties using only reflected signals constitutes the GPR inverse problem. Some of the methods applicable to GPR inverse problems are discussed in this chapter.

The sections that follow review the formulation of GPR inverse problems remarking on the complexity of reaching solutions from purely mathematical procedures. Alternative methods based on soft-computing and other artificial-intelligence-aided methods have been applied to assess the geometric and EM properties of subsurface targets. These are discussed briefly. This is followed by a review of the main steps in pattern recognition and the available tools and methods for the task. Although an exhaustive description of useful GPR cases is not feasible in the context of this chapter, successful examples of GPR prediction are included to illustrate the methodology and objectives at each stage. It is hoped that this can serve as a step-by-step guide to extension of the methods discussed here to other applications in GPR.

6.2 Inverse problems

To predict the features of GPR scenarios from measurements, a model of the system under investigation and a mathematical theory linking the physical description of the testing environment to the measurements are required. Previous chapters (see Chapters 2 and 5) have presented the equations and methods needed for accurate simulations of realistic GPR scenarios, needed to obtain the EM scattered fields. This constitutes the so-called forward problem, as depicted in Figure 6.1. The inverse problem consists in using a limited set of actual measurements to infer the values of the physical parameters of the GPR test environment.

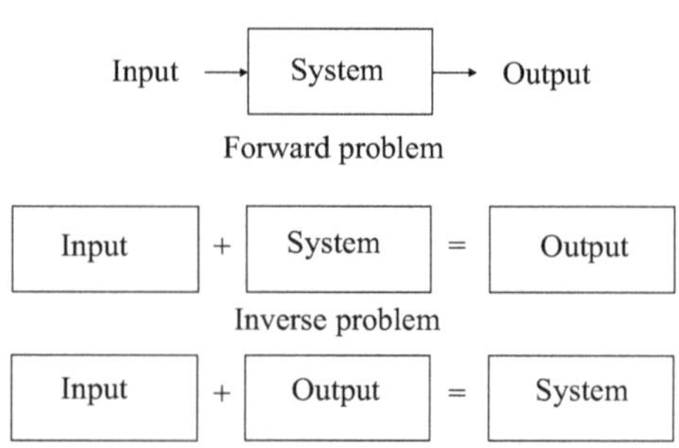

Figure 6.1 Schematic representation of the forward and inverse problems

Even a cursory inspection of the differential and integral equations on which the various models are based is sufficient to realize the challenges of formulating a mathematical description of inverse problems. For instance, for the simplest of problems, that of a homogeneous background medium, it is common to apply the Born approximation to linearize the problem and to derive closed-form relations between the EM incident field $\mathbf{E}_{inc}$ from the GPR transmitter and the scattered field $\mathbf{E}_s$ from buried targets. Under these conditions, the scattered field at any point r_0 at a particular frequency ω is

$$\mathbf{E}_s(\mathbf{r}_0, \omega) = k_b^2 \int_V G(\mathbf{r}_0 - \mathbf{r}', \omega)\mathbf{E}_{inc}(\mathbf{r}_0, \mathbf{r}', \omega)\chi(\mathbf{r}')d\mathbf{r}' \tag{6.1}$$

where G denotes Green's function, k_b is the wavenumber in the host medium, and the unknown contrast function χ is defined as the relative difference between the permittivity of the target and the reference homogeneous medium. For simple media (linear, isotropic, homogeneous) and simple inclusions, it is possible to find an appropriate Green's function and the problem may be solved analytically. Unfortunately, actual test environments are replete with difficulties, including complex surfaces and boundaries, inhomogeneous materials, and complex-shaped targets, for which the contrast function cannot be known. Noise, which is always present, only adds to the difficulties.

As a result, inverse problems involving EM wave propagation are ill posed. In general, any mathematical problem is called ill posed if one of the following three conditions is present:

1. **Nonexistence**. A model is assumed to exist but in fact a model that exactly fits the output data may not exist. Possible reasons for the nonexistence condition include inappropriate physical description (for instance, due to approximations to the exact mathematical equations), or too low a signal-to-noise ratio (SNR) in the output data.
2. **Nonuniqueness**. If exact solutions do exist, they may not be unique even for an arbitrarily large number of exact data points. That is, there may be other solutions aside from the one found that exactly satisfy the expected input. At this point, it is worth recalling the uniqueness theorem of EM fields, which relates unique solutions for a finite region to the knowledge of the tangential fields at the boundaries of the region. In that respect, measurements at GPR receiving antennas, even for arrays of receivers, constitute only a small portion of the tangential scattered fields and these may be distorted after propagation through various media.
3. **Instability**. The process of computing an inverse solution is considered unstable if a small change in measurements or outputs may lead to a large change in the physical description of the input problem. For instance, the estimation of characteristic inclusions (such as χ in (6.1)) can vary significantly when particular measurements with strong backscattered fields are added to the set of outputs. In those cases, more available data do not imply higher accuracy, and solutions often require a reevaluation of the relations between input and outputs.

The inverse problem in GPR is usually that of finding the characteristics of a target buried in a lossy dielectric medium. Outputs are the reflected fields measured by the antenna, normally as a collection of discrete observations in time. The parameters needed to characterize inclusions in the soil or other media (which is typically modeled as a dielectric slab) are the electric properties (permittivity and conductivity) and the geometrical characteristics (depth and dimensions such as radii of inclusions). When a discrete set of outputs is considered, the inverse problem is called a discrete inverse problem or parameter estimation problem.

One possible approach to solve the inverse problem is the use of parametric algorithms that, in turn, are based on mathematical optimization algorithms. This type of algorithms updates the parameters iteratively to minimize (or maximize) a certain evaluation function. However, that requires a forward solver in the optimization loop. The computation of approximate electric or magnetic fields can be performed by the finite-difference time-domain (FDTD) method, or any other numerical method (see discussion in Chapter 5), as a forward solver. The optimization process can be carried out by using a variety of algorithms such as the Newton method, conjugate gradient method, or the more recent evolutionary algorithms. A major limitation of using parametric algorithms is the time of calculation that can be prohibitive for complex 3D problem, because the forward solver must be executed many times before an acceptable solution is obtained.

Nonparametric algorithms can also be used. Their implementation is more complex, but they provide faster solutions because iterative calls to forward solvers are not required. A remarkable type of nonparametric methods is the class of migration algorithms, presented in Chapter 4 and originally used for seismic prospecting. The algorithms are not specifically linked to any discipline and can be applied in general media and various targets as will be discussed in the following subsection. A second kind of nonparametric technique is based on imaging algorithms and other pattern recognition methods. In this subgroup, a number of inversion methods have been developed to reconstruct the scattered signals back to their true spatial location, most of them based on the numerical inversion of integral equations [1,2]. These methods are characterized by a high level of complexity, accuracy, and as a result, a significant computational burden. In common with other inversion techniques, the imaging of typical GPR field data using these methods may be difficult due to problems of limited coverage, noisy data, or nonlinear relations between observed quantities and physical parameters to be reconstructed. As a result, it has become necessary to develop specific methods for interpretation of raw GPR data that address the problems encountered in GPR surveys. This type of analysis requires algorithms through which problems with complex scattering properties can be solved as accurately and as fast as possible. This is a difficult requirement to satisfy when dealing with iterative solution algorithms categorized by a forward solver as part of the loop. Even if the forward solver is only executed a limited number of times, the solution process becomes computationally unreasonable for large field-data sets. The following sections address some of the more common methods for inversion of GPR data and, in the process, discuss issues of feasibility and efficiency of methods.

6.2.1 Reverse-time migration algorithm

The migration process attempts to reconstruct the buried target geometry from the signals obtained at the surface. The technique has been developed primarily in acoustic, seismic, and geophysical engineering and was originally developed in two-dimensional form by Hagedoorn in 1954 [3]. Additional developments may be found in [4–6]. In [7], a pair of FDTD reverse-time migration algorithms were presented for radar data processing that uses linear inverse scattering theory to develop a matched-filter response for the radar problem. These algorithms were developed for both bistatic and monostatic antenna configurations.

Spatial location and reflectivity are the typical information obtained from radar. Since most radars use broad beamwidth antennas, the power reflected from a buried structure is recorded over a large lateral aperture in the image space. For example, a monostatic survey can be used to collect data over discrete objects such as a pipe, in which the diffraction will appear as a hyperbola in the space–time image (see, e.g., Figures 5.10 and 5.12). In this case, no further processing may be needed if the goal is simply to detect the pipe. However, in most cases, pattern recognition algorithms must be used to move the observed scattering events to their true spatial location and to estimate the target's reflectivity.

The objective of the migration algorithm is to calculate a mapping function $S(\mathbf{r})$, at any location on the surface, with values related to the local permittivity $e(\mathbf{r})$. The algorithm developed in [7] is based on the notion of a matched filter, which is used extensively in radar applications. The matched filter concept can be explained as a correlation of the received signal with the expected or estimated signal from a specific target. If this correlation produces a large value, then it is likely that the target is present. The resulting algorithm can be directly related to reverse-time migration. Using this development, an image can be perceived as a back-propagated wave-field reconstruction of the dielectric contrast within the host medium.

Mathematically, the matched filter transfer function H is expressed as the complex conjugate of the expected received waveform due to the target to which the filter is being matched. The output of the matched filter for N transmitters, located at the position vectors $\mathbf{r}_n''$, and an M-element receiver array, located at the position vectors $\mathbf{r}_m$ is expressed as [7]

$$S\left(\mathbf{r}'\right) = \sum_{n=1}^{N} \int \sum_{m=1}^{M} H\left(\mathbf{r}_n'', \mathbf{r}', \mathbf{r}_m; \omega\right) R_n\left(\mathbf{r}_m; \omega\right) d\omega \tag{6.2}$$

where R_n is the received waveform due to the nth transmitter as shown in Figure 6.2.

The problem geometry depicted in Figure 6.2 consists of two half-spaces where Region 1 corresponds to free space, whereas an inhomogeneous ground, characterized by constitutive parameters μ_0, $\varepsilon_2(r)$, is denoted as Region 2. Considering a weakly scattering object of finite size with constitutive parameters μ_0, $\varepsilon(r)$ located within the ground, the matched filter will maximize the output power at $t = 0$ if the complex conjugate of the matched filter output is equal to the received signal.

The electric field at any location $\mathbf{r}$, scattered by the buries object can be expressed as

$$\mathbf{E}_{sca}\left(\mathbf{r}\right) = -\int \overline{\mathbf{G}}\left(\mathbf{r}, \mathbf{r}'\right)\left[k^2\left(\mathbf{r}'\right) - k_b^2\left(\mathbf{r}'\right)\right]\mathbf{E}_{inc}\left(\mathbf{r}'\right) d\mathbf{r} \tag{6.3}$$

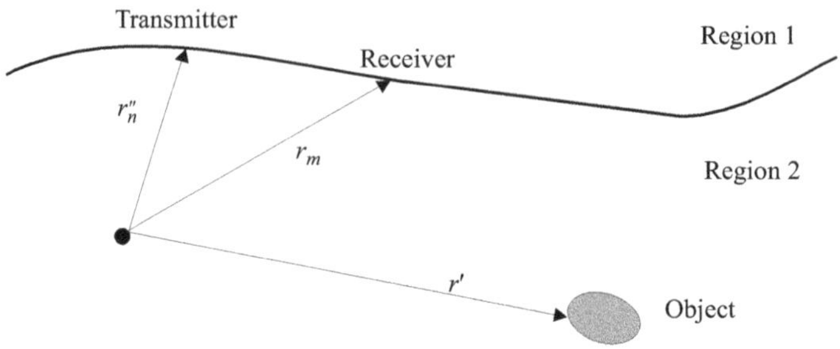

Figure 6.2 Geometry of a generic radar problem [7]

where $\mathbf{E}_{inc}(\mathbf{r}')$ indicates the incident field at the scatterer and $\overline{\mathbf{G}}(\mathbf{r},\mathbf{r}')$ is the background dyadic Green's function, and k and k_b correspond to the wavenumbers for the cases without and with the scattering object present, respectively. If the scatterer is small, and normalizing by $\varepsilon(\mathbf{r}) - \varepsilon_2(\mathbf{r})$, the following equation holds approximately

$$\mathbf{E}_{sca}(\mathbf{r}) = -\omega^2\mu_0\overline{\mathbf{G}}(\mathbf{r},\mathbf{r}')\,\mathbf{E}_{inc}(\mathbf{r}') \tag{6.4}$$

Additionally, it can should be noted that the waveform at the receiver can be calculated as $\mathbf{u}_r^* \cdot \mathbf{E}_{sca}(\mathbf{r})$, and that any signal T radiated by the transmitting antenna reaches the scatterer as $\mathbf{E}_{inc}(\mathbf{r}') = j\omega\mu_0\bar{\mathbf{G}}(\mathbf{r}',\mathbf{r}'')T\mathbf{u}_t$, where $\mathbf{u}_r$, $\mathbf{u}_t$ are the effective lengths of the receiver and transmitter antennas, respectively and * indicates the complex conjugate. By using this approximation, the matched filter H can be written as

$$H(\mathbf{r}'',\mathbf{r}',\mathbf{r}) = \mathbf{u}_r^*\left\{-\omega^2\mu_0\overline{\mathbf{G}}(\mathbf{r},\mathbf{r}')j\omega\mu_0\overline{\mathbf{G}}(\mathbf{r}',\mathbf{r}'')\,T\mathbf{u}_t\right\} \tag{6.5}$$

The final image is then expressed by applying the complex conjugate of the measured data to the filter

$$S(\mathbf{r}') = \left\{j\omega\mu_0\overline{\mathbf{G}}(\mathbf{r}',\mathbf{r})\,[-j\omega R\mathbf{u}_r]^*\right\}\left\{j\omega\mu_0\overline{\mathbf{G}}(\mathbf{r}',\mathbf{r}'')\,T\mathbf{u}_t\right\} \tag{6.6}$$

Equation (6.6) gives the migrated data as a function of frequency and the transmit and receive antenna locations. The first term is the electric field intensity generated by a current source $[-j\omega R\mathbf{u}_r]^*$. If the time dependency of the received signal is introduced, the source is expressed as a derivative and time reversal $[-j\omega R\mathbf{u}_r]^* \Rightarrow R'(-t)\,\mathbf{u}_r$. The field generated by this source will be referred to as the back-propagated electric field $\mathbf{E}_{bp}$. The second term is simply the incident field intensity $\mathbf{E}_{inc}$. Reintroducing the frequency dependencies and referring to (6.2), a complete expression for the migrated data is now available as

$$S(\mathbf{r}') = \sum_{n=1}^{N}\int\sum_{m=1}^{M}\mathbf{E}_{mn,bp}(\mathbf{r}';\omega)\,\mathbf{E}_{n,inc}(\mathbf{r}';\omega)\,d\omega \tag{6.7}$$

$$\mathbf{E}_{mn,bp}(\mathbf{r}';\omega) = j\omega\mu_0\overline{\mathbf{G}}(\mathbf{r}_m,\mathbf{r}';\omega)\,[-j\omega R_n(\omega)\,\mathbf{u}_r]^* \tag{6.8}$$

$$\mathbf{E}_{n,inc}(\mathbf{r}';\omega) = j\omega\mu_0\overline{\mathbf{G}}(\mathbf{r}',\mathbf{r}_n'';\omega)\,T(\omega)\,\mathbf{u}_t \tag{6.9}$$

where the subscripts *m* and *n* denote the field or signal due to the *n*th transmitter and *m*th receiver.

6.2.2 Pattern recognition algorithms (PRAs)

Image and pattern recognition techniques have been an active line of research in the last 50 years, moving from basic problems related to classification of photographs and satellite images to more challenging realistic problems such as forensics, real-time videos, airborne radars, and, of course, GPRs. Regardless of the specific application, the methods are characterized by a two- or three-stage process. Starting from a set of, usually large, raw data, the objective is to provide some classification of the problem under consideration. This classification can be either qualitative (for instance, the presence/absence of any buried scatterers in a survey) or quantitative (e.g., any measurable quantities such as the electrical permittivity of buried scatterers or that of a location in that ground with varying properties). In a two-stage algorithm, raw data are processed directly in a classifier by the so-called end-to-end deep learning techniques (these include auto-encoders, Boltzmann machines, convolutional and recurrent NNs, and so on). On the other hand, in three-stage algorithms, raw data are preprocessed to form a feature space before invoking the classifier. If the domain-specific feature representation is judiciously chosen, the dimension of the raw data is reduced to such a level that basic statistical pattern recognition classifiers provide computationally fast results. The success of these techniques then relies on both the initial preprocessing algorithms (including signal filtering, variance analysis, and feature extraction), and the subsequent pattern recognition methods (including feature selection, Bayesian theory, NNs, ensemble methods, clustering, and many more). The list of the potential useful methods for GPR is extensive, and algorithms can also be applied concurrently searching for improved results in both two- and three-stage processes. In practice, any pattern recognition algorithm could be successful for a particular GPR case as long as it is robust. Unfortunately, robustness in pattern recognition for GPR data is a problem yet to be resolved. Nevertheless, GPR applications benefit from these methods even if absolute robustness cannot be guaranteed in general.

Whereas analytical and simulated configurations can often be shown to be robust, and pattern recognition algorithms have been shown to be successful in controlled experiments, there are three basic assumptions that have to be removed to assure robustness in general, practical applications. First, the methods must be able to operate beyond any restrictive assumptions and must be capable of operating in environments and with values that were not considered in the design of the algorithms. That is, the methods must be adaptive. An example of this in GPR might be the correct assessment of the properties of inclusions in the presence of magnetic materials. Second, the methods must not rely on uniform and independent distributions. Material properties cannot, in general, be assumed to be uniform (e.g., permittivity and conductivity of soils) or independent of other parameters (such as depth). The algorithms must be able to identify new features or extend the set of objective parameters without significant changes in their original design. Finally, the assumption that data are "clean" and sufficient limits the capability of algorithms to produce valid results. Algorithms must be designed to operate with limited data and in the presence of noise and clutter, conditions that are characteristic of GPR surveys.

A schematic representation of the basic tasks performed by a typical three-stage imaging and diagnosis algorithm for GPR is shown in Figure 6.3. The steps are as follows:

- **Preprocessing**. The step is intended to reduce noise and undesired EM system effects (signal processing).
- **Image segmentation**. Done with an artificial NN (ANN) to classify areas potentially containing object reflections.
- **Pattern recognition**. Subsequent diagnosis by machine learning (ML) algorithms to identify patterns.

In the first step, the preprocessing procedure is called upon to reduce noise, to eliminate the undesired presence of the ground surface echo and to compensate for EM propagation losses [8]. Computational resources and effort are required in analyzing the data collected before processing, focusing on the main requirements, expected outcomes, and to set the most effective strategy for reaching the surveys' objective. The quality of the dataset influences the extent of the processing needed in order to increase data readability. At the same time, the level of noise and the ever present compromise between frequency, resolution, and range, in line with the nature of the target, may also represent a limit to the performance of the processing phase.

In the second stage, an image segmentation is performed. The use of convolutional NN (CNN or ConvNet) is often the method of choice. The CNN is a class of deep, feed-forward ANNs [9]. A review of object detection based on CNN can be

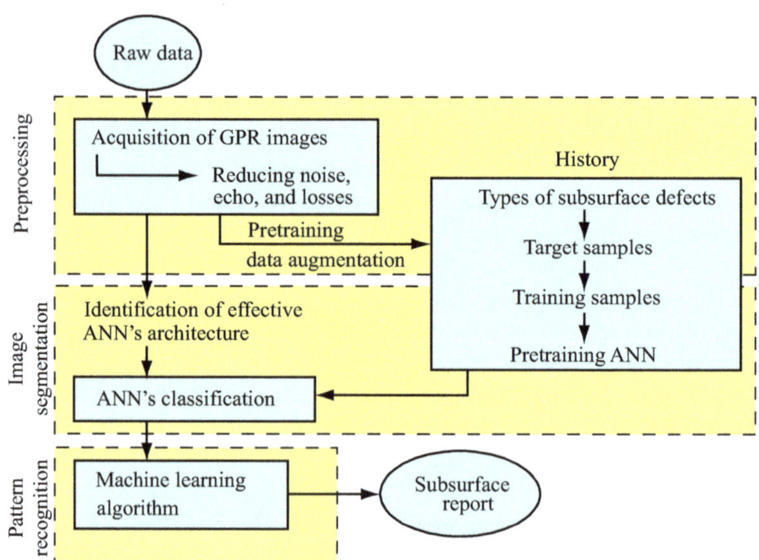

Figure 6.3 Main stages of GPR imaging and diagnosis system

found in [10]. The objective is to generate a map of all detected areas in the image. In effect, every pixel needs to have a label associated with it so that the last fully connected layer of the network is replaced by another convolution layer with a large reception field. The idea is to capture the global context of the image. CNNs are almost always used because they outperform other machine learning algorithms for image segmentation [10,11]. The benefit of using a CNN is that features extracted from the data are a learned parameter of the system, and the multilayer structure can be used to achieve minimal preprocessing. Some of the key user interventions that can improve the performance and the implementation of CNN for GPR data are (see [11]) as follows:

- use smaller convolutional filters to reduce the number of parameters needed to be learned;
- use network pretraining to increase the effective number of training data;
- augment the training dataset with data transformations, such as scale and rotation; and
- identify the best network architecture for the GPR problem under consideration.

Once images are properly segmented, pattern recognition techniques become effective. Support vector machine (SVM), hidden Markov model (HMM), k-NN, multi-class support vector machine (MC-SVM), Decision tree (DT), gradient-boosted decision tree (GBDT), adaptive boosted decision tree (ABDT), and other machine learning algorithms have been used as pattern recognition techniques to identify the hyperbolic anomalies associated with buried targets, generating information for decision-making actions [12–14]. Strengths and weaknesses of applying each technique in relation to GPR data are sometimes quite subtle, as can be seen in [15–20].

ANNs have been proven to be consistent options [21,22] with a track record that goes well beyond GPR. ANN-based algorithms learn (i.e., progressively improve performance) by incorporating examples, generally without task-specific programming. Methods of transcribing and transforming GPR images into formats useful for assessments can be found elsewhere [23]. A review of fault and error tolerance in ANN is available in [24] where the main passive techniques used to improve the fault tolerance of ANNs are presented. Other techniques for image recognition and classification have been used, regardless of their association with ANNs.

Machine language (ML) algorithms evolved from the study of computational learning theory in artificial intelligence [25]. ML algorithms are constructed in such a way as to learn and make predictions from the available data, unlike static programming algorithms that need explicit human instructions. Many consider artificial neural networks as a machine language technique because of the learning property inherent in ANNs. There are other views on the subject [26]. For instance, some view ML as "regression and classification" algorithms and ANN as "deep learning" for multilayer perceptrons (MLPs), CNNs, cost functions, and back-propagation algorithms. Recent studies provide a general comparison between a number of different ML algorithms [27,28]. Theoretical and experimental data-modeling is discussed in [29] in large-scale data-intensive fields, relating to model efficiency, including computational requirements in learning. For the purpose of this chapter, a pragmatic view is more appropriate, whereby ANN and ML are viewed as two different categories of algorithms and suggest the use of ANN for image segmentation (to classify areas potentially containing object reflections) and after diagnosis use ML techniques to identify patterns.

To understand the importance and the current state of pattern recognition algorithms in connection to GPR, it is useful to look at a historical timeline of contributions to the field. The first published application of ANN to GPR data may be found in [30]. The initial method consisted of an algorithm that adapts the input data to search objects' signatures that are successively identified by means of a recognition step using a back-propagation NN. It identified buried pipe signatures with a degree of accuracy similar to those performed by a human operator. ANNs, machine language, and related methods have since evolved considerably as can be seen in Figure 6.4 and Table 6.1. Figure 6.4 shows the number of publications per year over the period 2000 and 2018 as well as the time of emergence of various methods of analysis. Table 6.1 indicates the main contributions to the state of the art during the same period. The progression includes applications related to (primarily) engineering and computer science (47%), physics and astronomy (13%), materials science (12%), and earth and planetary sciences (11%). The timeline suggests that ANNs were first used to identify hyperbolic signatures in the late 1990s to the early 2000s. The procedure then evolved to the use of fuzzy logic methods for the detection and localization of targets. That period corresponds to the emergence of GPR as an NDE technique.

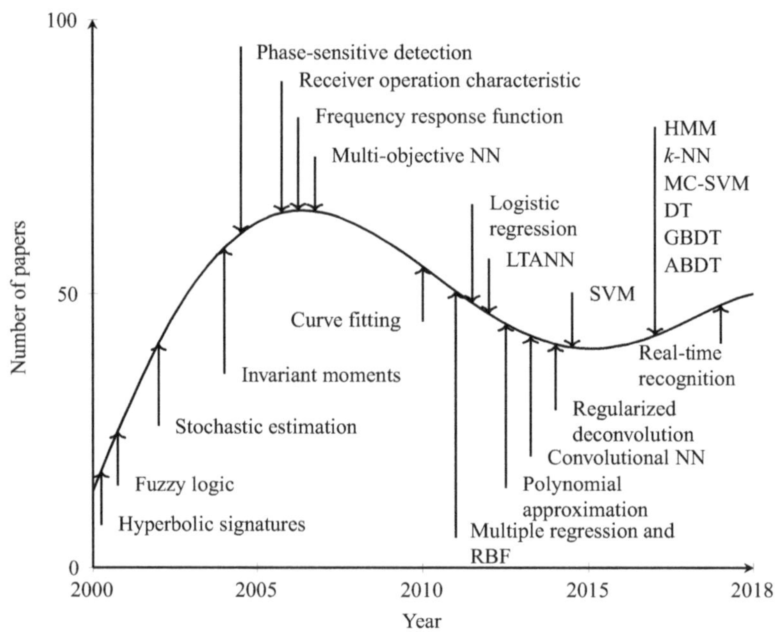

Figure 6.4 A timeline of emergence of analysis methods and rate of publications on GPR

Table 6.1 Main contributions to analysis of GPR data based on ANN and ML methods

Reference	Main contribution
[34]	ANN performs with the same accuracy as a human operator.
[35]	Detection algorithms are combined using fuzzy logic and ANNs.
[36]	ANN operates with 3D images to locate targets.
[37]	Stochastic estimation is combined with ANNs.
[38]	Invariant moments as a feature extraction for ANNs.
[39,40]	ANN-based reconstruction based on EM-fields.
[41]	Receiver operating characteristic (ROC) curves to train ANNs.
[42]	Principal component analysis (PCA) to train ANNs.
[43]	Frequency response function (FRF) and ANN.
[44]	Multi-objective neural networks (MNNs).
[45]	ANNs and curve fitting techniques.
[46]	Multiple regression and the radial basis function neural network.
[47]	ANNs and logistic regression algorithms.
[48]	Laplace transform instead of ordinary weights (LTANN).
[49]	Polynomial approximations (the feature vector) are ANN inputs.
[31,50]	The benefit of using a convolutional NN (CNNs) is that features extracted from the data are a learned parameter of the system.
[51]	Regularized deconvolution is utilized to increase range resolution.
[52]	CNNs configuration to extract data from multiple sources.
[12–14]	ML approaches are used to detect signatures with ANNs.
[53]	Real-time hyperbolae recognition and fitting in GPR data.
[54]	A multitask spatiotemporal ANNs that combine 3D ConvNets and recurrent neural networks (RNNs).

As more applications were tested, more complex the data became with a higher false alarm rate. More robust algorithms were necessary to cope with this issue. During the 2000s, one can clearly see a shift toward stochastic methods.

An important subject in conjunction with ANN and/or ML techniques for GPR is the training data. Historically, most training data came from a relatively limited set of measured raw data. In the current environment of faster full-wave forward solvers and improved analytical methods, it is pertinent to discuss the use of simulated results as training data, their value, and their relation to real data. Simulated data of realistic test geometries can offer the possibility of providing a more diverse set of training data, than solely measured results. For example, the use of CNNs was mentioned earlier as an advantageous method because features extracted from the data are a learned parameter of the system. But, at the same time, a well-known limitation of CNNs is that they require large amounts of data for training (parameter inference) to avoid overfitting (poor generalization). One approach to alleviate this difficulty is the use of pretraining data [31]. Pretraining can lead to improved detection performance but some configurations also cause a deterioration in performance. Although pretraining is a common strategy, it is not a guarantee for improvement, and both the issue of pretraining and the volume and diversity of training data necessary to obtain robust ANN and ML performance have received considerable attention. For example, techniques for training data augmentation [11,32,33] and methods of initialization of the pretraining steps indicate that both pretraining and dataset augmentation can help one to achieve higher GPR detection performance. Many of these needs can be met with simulated data provided of course that the data are realistic.

It is also important to discuss restrictions on machine learning (and ANN) methods to specific application areas such as rebar diameter estimation, cylinder and pipeline location, or detection of small asymmetric cracks. The measures of "condition positive," "condition negative," "true positives," "true negatives," "false positives," and "false negatives" are measures commonly used to estimate the accuracy of the classifier. For example, criteria and tests for ML techniques applied to GPR detection of landmines include misclassification rate (indicates the probability of the classifier being wrong); true-positive rate (indicates how often it predicts positive, also known as sensitivity or recall); false-positive rate (indicates how often it predicts positive when it is negative); specificity (indicates how often it predicts negative when it is negative); precision (indicates the classifier's correctness when it predicts positive); false-negative (indicates the classifier's incorrectness when it predicts negative); F score (this is a weighted average of the true-positive rate and precision); and safety factor (it is the probability of a mine being detected when is not a mine) [14]. The success rate of any GPR system is limited by the sensing aspects of each individual application and tends to produce high false alarm rates when the situation is generalized.

6.3 Pattern recognition methods applied to GPR

Some examples of applications designed to assist the operator in the interpretation of results of GPR surveys are described next starting with a parametric technique (model fitting) based on optimization. As was mentioned earlier in this chapter, nonparametric methods are currently more commonly employed because of the lower computational resources required. For this reason, a comparison of results for a hybrid model fitting and migration techniques for the same example is useful and hence discussed here. Although iterative optimization approaches such as those considered here take considerable computation time, these approaches provide much better image qualities for high-contrast objects than other linear inversion methods such as diffraction tomography [55,56]. Taking into account the number of contributions to GPR pattern recognition, ANNs are presently studied more often than other methods; therefore, a practical case that addresses ANNs is included for completeness. To illustrate the complexity of predicting multiple geometric/electric properties of inclusions in GPR scenarios, a multi-objective ANN is also briefly described. Finally, and also belonging to the group of nonparametric techniques, the feature selection approach is analyzed in a separate subsection to illustrate the relevance of the definition of feature space in a three-stage pattern recognition algorithm to provide a fair comparison with other commonly used PRAs.

6.3.1 Buried cylinders in nonhomogeneous dielectric media: model fitting and hybrid migration-model fitting approaches

This first example combines two algorithms for processing GPR 2D data by means of linear inverse scattering. The GPR problem consists of locating and characterizing a finite number of parameters of cylindrical inclusions buried in a nonhomogeneous dielectric host medium. The properties addressed are the electrical permittivity and conductivity and depth and radii of the inclusions. The configuration is described in Figure 6.5. It consists of a set of inclusions with different properties and dimensions at various depths. The goal is to identify and characterize the geometry (radii and depths) and electric properties (permittivity and conductivity) of the inclusions from the scattered field.

The dielectric medium in this example is simulated with electrical characteristics of concrete [57] with a mean relative electrical permittivity value of 6 and standard deviation 0.25 (thus mimicking a nonhomogeneous medium). An FDTD forward solver is used, with a pulse source defined by a differentiated Gaussian pulse with a center frequency $f = 900$ MHz and bandwidth between 0.3 and 2 GHz (Figure 6.6). The GPR antennas are simulated as line currents of infinite extension in the z-direction (2D problem). A sampling time interval of 20 ps was chosen to meet the Courant stability criterion of the FDTD method, with 1,500 samples collected for each trace, corresponding to a time interval of 30 ns. In order to control the numerical dispersion and provide a good discretization for the inclusions, the spatial steps were chosen as $\Delta x = \Delta y = 12$ mm. The goal of this problem is, given the incident and scattered waves, to determine the physical and spatial characteristics of the inclusions. The raw EM field data are collected at a fixed number of receiver locations (in this example, 31) distributed uniformly (7.2 cm apart) relative to the transmitter, which is centered

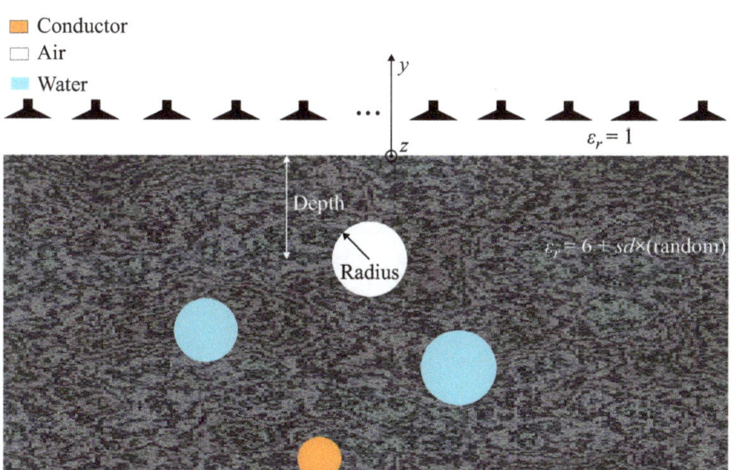

Figure 6.5 Configuration consisting of circular cylinders located in a nonhomogeneous dielectric

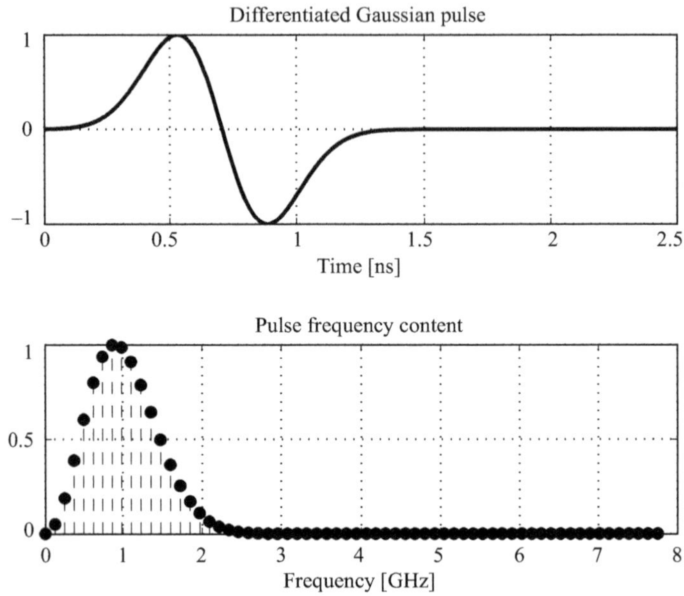

Figure 6.6 Pulse description

above the dielectric medium. Increasing the number of receivers does not necessarily improve the prediction, not only because it increases the computational time for the forward solver, but also because it can drive the pattern recognition algorithm to instability.

The performance of imaging algorithms is considered first, by applying the parametric model fitting method. In this method, target characteristics and location are expressed as varying input parameters. These parameters are updated iteratively to minimize, by means of optimization algorithms, the differences between the observed data and the simulated data. Metrics to evaluate this difference are based on evaluation or cost functions, defined in terms of scattered fields (measurements) for the reference configuration, $\mathbf{E}(\theta_0)$, and the scattered fields (simulations) for a test configuration $\mathbf{E}(\theta)$ [58]. The solution is the optimal θ^* that minimizes the error between the reference and test configurations. Mathematically

$$\theta^* = \arg\min f(\theta) = \sum_{i=1}^{N_s} |\mathbf{E}(\theta_0) - \mathbf{E}(\theta)|^2 \tag{6.10}$$

where N_s is the number of sample points at which the scattered wave is measured. Note that $\mathbf{E}(\theta_0)$ is known (measured) even though the actual configuration θ_0 is unknown. The scattered field $\mathbf{E}(\theta)$ is iteratively generated assuming configuration tests θ, and the optimization procedure attempts to minimize the error between $\mathbf{E}(\theta_0)$ and $\mathbf{E}(\theta)$ in such a way as to identify the actual configuration θ_0, that is, at some level of error, one can assume that the simulated configuration is sufficiently close to the reference configuration.

It is important to remark that the problem described in (6.10) is usually multimodal. This can be seen in Figure 6.7 that shows a simplified case in which the unknowns are the radii of two inclusions (their physical properties were fixed and known, to reduce the dimensionality of the problem). Even in this simple example, the surface is clearly multimodal. Deterministic mathematical optimization algorithms perform poorly for multimodal problems because the optimization often fails, trapped in a local minimum. Stochastic approaches are preferred to cover large search spaces. In this case, the particle swarm optimization (PSO) method [59] is used, even though other stochastic algorithms can be successfully applied. PSO is one of the latest evolutionary optimization methods inspired by nature, a group that includes other evolutionary strategies such as genetic algorithm or ant-colony optimization. PSO is based on the metaphor of social interaction and communication such as bird flocking and fish schooling. PSO is distinctly different from other evolutionary-type methods in that it does not use a filtering operation (such as crossover and/or mutation) between generations, and the members of the entire population are maintained throughout the search procedure so that information is shared among individuals to direct the search toward the best position in the search space. However, regardless of the specific stochastic algorithm employed, the number of forward simulations required to optimize the cost function grows drastically when the number of unknown parameters (i.e., the dimension of the search space) increases.

Nonparametric algorithms can be equally well used to solve this problem. Usually these algorithms are more complex to implement but can solve inverse problems

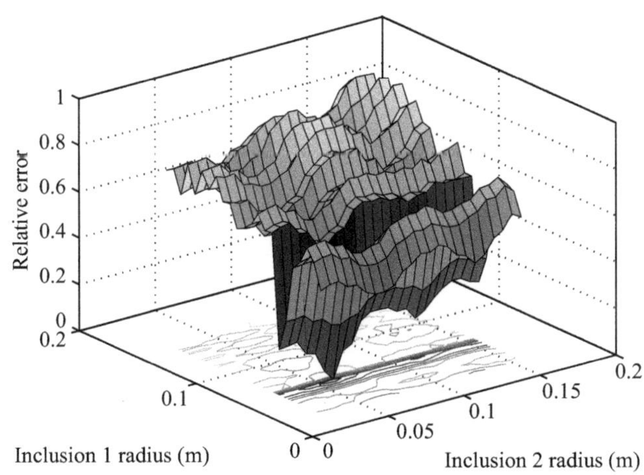

Figure 6.7 Surface plot of (6.10) representing the problem of finding the radius of two inclusions, given that their location and physical properties are known

faster than parametric algorithms because they do not rely on iterative evaluations of the function. In this example, a matched-filter-based reverse-time (MFBRT) migration algorithm (designed for waves in general vector fields) was implemented on the basis of the idea of a matched-filter [7]. Even though this is a robust method for solving the abovementioned problem, it can only deal with the number of inclusions and their centers, whereas other electric and geometric properties are out of scope for this approach. The implementation in FDTD is accomplished by propagating the incident field in reverse while simultaneously propagating the back-propagated field forward. The matched filter uses a limited set of simulations in a known homogeneous medium to locate the inclusions, and the solution is taken as that at which the incident field intersects the back-propagated field. As an example, in the GPR scenario in Figure 6.5, the four cylinders (with different radii), and three possible electrical properties (conductor (1), air (2) and water (3)) are buried in concrete. Figure 6.8 shows the final image obtained at $t = 0$. The image provides a more accurate location of the inclusion positions as opposed to, for example, Figures 5.10 and 5.12 in which only the general position can be ascertained. It can be noticed that this process improves the final images compared to any direct inspection of the raw data.

There are, however, some major drawbacks in this approach. First, it assumes that the background medium is known (this is a common assumption in many commercial software tools for the interpretation of GPR data [60]). In addition, the discrimination of targets in close proximity is somewhat limited. And, finally, it cannot provide further characteristics of the inclusions, such as radii and none of the EM properties.

For this reason, a combination of the MFBRT and PSO algorithms is proposed to take advantage of their main strengths and to minimize their shortcomings. The MFBRT alone would not be capable of delivering a complete description of the

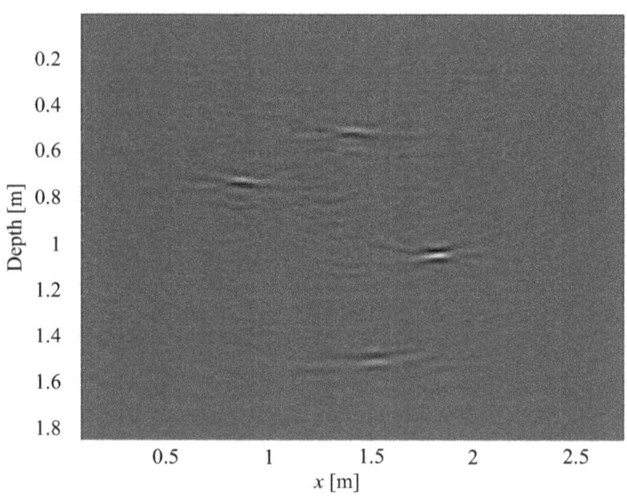

Figure 6.8 Bistatic reverse-time migration for the geometry in Figure 6.5; final data

inclusions, but it helps in decreasing the PSO computational burden as the number of optimization variables is reduced. The proposed hybrid algorithm was tested for a simple case of inclusions defined in Table 6.2, given a non-homogenous medium with a fixed standard deviation ($sd = 0.25$). In the hybrid MFBRT-PSO algorithm, the search space assumes that the radii are in the range of 5–10 cm for each inclusion, and the constituent material belongs to the set {air, conductor, and water}.

The hybrid approach that combines the MFBRT with PSO is compared with the PSO alone considering problems with one, two, and three inclusions in Table 6.3 in terms of the number of function evaluations. As with all stochastic algorithms, the number of simulations is taken as an average of the required evaluations to properly define the inclusions for ten different executions. The results show that the number of function evaluations decreases drastically when the hybrid approach is applied. For the case of PSO alone, the number of evaluations diverges for as few as three inclusions. However, the use of the hybrid approach in situations where the number of inclusions is high is also limited, since the number of evaluations increases rapidly.

In summary, these examples illustrate the strengths and weaknesses of parametric and nonparametric approaches for the inverse scattering problem in non-homogenous host media. The nonparametric approach is a robust and fast technique capable of correctly finding the number of inclusions and their centers but incapable of defining their geometry and physical properties. The parametric approach can solve the general problem but is more expensive in computational terms. As well, stochastic algorithms are an absolute necessity because the estimation of inclusions in a large

Table 6.2 Definition of inclusions for inverse scattering calculations

	Center (x,y) [m]	**Radii [cm]**	**Material**
Inc_1	(0.78,0.84)	6.6	Air
Inc_2	(1.32,0.60)	9.0	Conductor
Inc_3	(1.74,0.60)	9.6	Water

Table 6.3 Number of FDTD evaluations spent on average, taking into account ten test cases, by the hybrid approach proposed in this work compared with the PSO alone

Number of inclusions	**Hybrid**	**PSO**
1	41	670
2	475	3,800
3	2,100	NA

search region with limited-view measurements often fails by settling on a local minimum. These two approaches are combined in such a way as to try to take advantage of their strengths to decrease the computational burden. This combination is done in two phases: (1) the nonparametric approach is first used to define the number of inclusions and their centers, (2) the parametric approach is then used to define the other unknown variables, the inclusions radii and the physical properties. This combination can perform characterization with a fraction of the function's evaluations necessary for the parametric approach used alone. This approach yields accurate detection of positions and effective physical properties of the targets.

6.3.2 Buried cylinder in nonhomogeneous dielectric medium: the artificial neural network approach

This section addresses the 2D GPR problem of a single cylinder of unknown characteristics buried in a nonhomogeneous dielectric. The aim is to accurately predict the depth and radius of the cylinder regardless of its electrical characteristics. In this case, by way of example, the ANN algorithm is applied as the pattern recognition algorithm. The main advantage of ANNs compared with the stochastic optimization approaches of the previous section is a lower computational effort for the evaluation process. The dielectric medium uses the electrical characteristics of concrete [61] with a mean relative electrical permittivity value of 6 and standard deviation 0.15, i.e., a mildly non-homogenous medium. The dielectric medium is illuminated with the differentiated Gaussian pulse in Figure 6.6 with a center frequency $f = 900$ MHz and bandwidth between 0.3 and 2 GHz. Antennas being simulated as current lines in an FDTD forward solver (Figure 6.9). The numerical dispersion and the quality of discretization are controlled by selecting the spatial step as $\Delta x = \Delta y = 6$ mm. Given an incident wave, W_i, and a scattered wave, W_s, the ANN determines the *radius* and *depth* of the inclusion.

ANNs are parallel computational models that comprise densely interconnected adaptive processing units called neurons. Neurons are usually defined to be similar

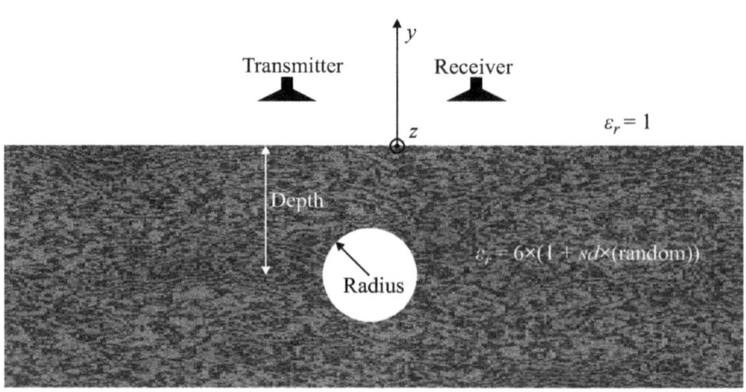

Figure 6.9 Buried cylinder in nonhomogeneous dielectric: problem description

and can be interconnected in various ways to create the network. The ANN achieves its ability to learn and then recall that learning through the weighted interconnections of the neurons. The interconnection architecture can vary for different types of networks. A very important feature of this methodology is its adaptive nature, whereby the problem is solved by feeding the system with examples akin to the way the human brain learns. This feature makes such computational models very appealing in application domains where one has little or incomplete understanding of the problem to be solved but where training data are readily available. Another key feature is the intrinsic parallel architecture that allows for fast computation of solutions. ANNs can be used in a wide range of applications, including pattern classification, speech synthesis and recognition, adaptive interfaces between humans and complex physical systems, function approximation, image compression, associative memory, clustering, forecasting and prediction, combinatorial optimization, nonlinear system modeling, and control. In the context of GPR, ANNs are used in the frame of the intelligent signal processing (ISP), a method characterized by the use of model-free (intelligent) methods based on training data. In addition, ISP implies the ability to extract system information from the example data alone and is less dependent on a priori environmental and system information or, stated simply, the method is less unstable.

The use of ANNs in GPR inverse scattering problems using parallel networks and networks with multiple outputs for a homogenous host medium was presented in [39, 62]. In [39], it is shown that both configurations can deliver reasonable and very similar results using input parameters from the scattered wave defined as (1) the peak amplitude of the reflected field; (2) the delay of the first reflected echo, calculated with respect to the time of arrival, at the receiving point; and (3) a measure of the duration of the scattered field (see Figure 6.10). However, these parameters are not sufficient to solve the same problem in a non-homogenous medium, and an algorithm to "squeeze" the scattered wave in order to collect more information by using PCA becomes necessary. The ANN used in this section is an asynchronous ANN called parallel layer perceptron (PLP) [63].

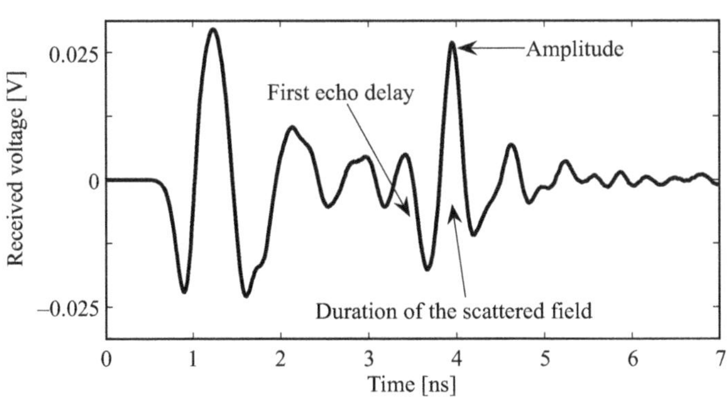

Figure 6.10 Reflected wavefield from a buried target

The PLP is a network that combines the advantages of both MLPs and adaptive-network-based fuzzy inference systems (ANFIS). The former has gained popularity in a vast range of applications due to its universal approximation characteristics. The latter has some advantage over MLPs due to the linear dependence of its output in relation of the result. However, the use of ANFIS is prohibitive for problems with several variables, such as is the case in almost any practical GPR problem. Therefore, the purpose of the PLP is to try to minimize these limitations by extending ANNs to parallel environments. The MLPs have achieved similar results in experiments but they require a longer training time [63], which limits the application of these architectures to more complex GPR cases.

The output y_t of the PLP with n inputs and m perceptrons per layer is calculated as

$$y_t = \beta\left(\left\{\sum_{j=1}^{m}\left[\gamma\left(a_{jt}\right)\phi\left(b_{jt}\right)\right]\right\}\right) \tag{6.11}$$

$$a_{jt} = \sum_{i=0}^{n} p_{ji}x_{it} \tag{6.12}$$

$$b_{jt} = \sum_{i=0}^{n} v_{ji}x_{it} \tag{6.13}$$

where $\beta(.)$, $\gamma(.)$, and $\phi(.)$ are activation functions (such as hyperbolic tangent, Gaussian, and so on), v_{ji} and p_{ji} are components of the weight matrices P and V, x_{it} is the ith input for the tth sample, where x_{0t} is the perceptron bias, and y_t is the tth position of the vector output y.

One particular case of the topology is obtained by considering both $\gamma(.)$ and $\beta(.)$ as identity functions. In this case, the network output is computed as

$$y_t = \sum_{j=1}^{m}\left[a_{jt}\phi(b_{jt})\right] \tag{6.14}$$

The particular case in (6.14) has interesting properties, as pointed out in [63], and will be adopted in the following numerical experiments. Notice that the number of free parameters p_{ji} and v_{ji} grows linearly with the dimension as can be observed from (6.12) and (6.13). The specific PLP should be chosen to minimize the set of possible functions $y = f(w,x)$ where w, usually $\in \Re$ [64], represent the discrete version of the functions $v(.)$, $p(.)$ and $\phi(.)$ which define the architecture of the NN in (6.14).

To find the w coefficients, a mathematical optimization problem is defined as

$$w^* = \arg\min_{w} R(f(w,x)) \tag{6.15}$$

where $R(.)$ is a predefined quality criterion and w^* is the argument that minimizes (6.15). To solve (6.15), it is required to define the best approximation of the desired output, d, as a measure of a discrepancy function, $L(.)$, between the desired and obtained outputs. Therefore, the expected risk (error) between the desired and the approximate outputs can be expressed as [64]

$$R(w) = \int L(d, f(x, w))\, dF(x, d) \tag{6.16}$$

In general, the integral in (6.16) cannot be evaluated directly since the distribution $F(x, d)$ is unknown and the only available information is provided by a training set. The training set $S = z_1, \ldots, z_T$ consists of T random vectors $z_i = (x_i, d_i), i = 1, \ldots, T$, independently and identically distributed according to some unknown but fixed probability distribution $F(z)$. The risk $R(.)$ defined in (6.16) can be asymptotically approximated, given some consistence conditions [64], provided the number of training set samples T tends to infinity. Of course, such an infinite size set is not available. To overcome this problem, resampling techniques can be used to approximate the expected risk with a minimal number of T samples. The simplest strategy is the hold-out resampling, also called external validation or, simply, validation. It consists of removing samples from the initial learning set, and using them for validation. A k-fold cross-validation consists in dividing the training data S into k sets of approximately the same size, in such a way that the learning takes place in $k - 1$ sets and the model is independently validated with the remaining set. This independence in the validation process prevents the use of the same structure in the training and validation processes. This is performed k times using all the k folds as validation sets once. Then, the expected risk is considered to be the mean of the risks evaluated for all folds. An illustration of this strategy is shown in Figure 6.11 for a 3-fold cross-validation.

For the problem of the cylinder buried in a non-homogeneous soil, the ANN is trained with a set of different inclusion examples, say $S(.)$, with different *radii*, *depths*, ε_r and σ, numerically generated and simulated by the FDTD forward solver. Specifically, the ANN has been trained with a set of 1640 different examples of inclusions constructed by varying the *radius* in the range $[0.02, 0.1]$ m according to the rule $radius = 0.02 + i \times 0.001$, $i = 0, \ldots, 80$, permittivity ε_r in the range

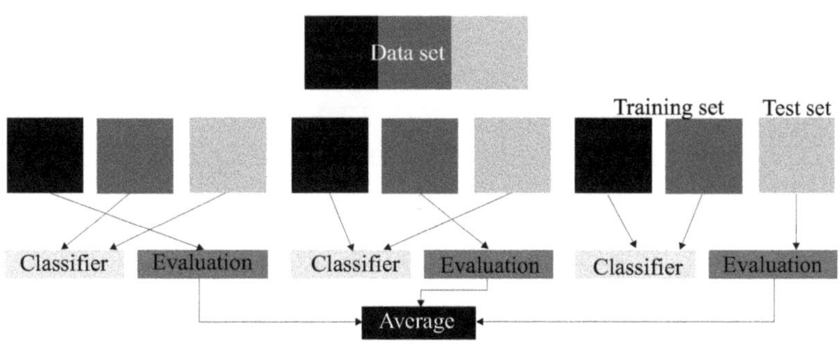

Figure 6.11 Example of 3-fold cross-validation. The two sets labeled as training sets are used to train the neural network and the error is evaluated on the test set. After all the folds are used once as test set, the expected risk is estimated as the average error of the sets

[1, 10], according to the rule $\varepsilon_r = 1 + i \times 1, i = 0, \ldots, 9$, conductivity σ in the range [0, 4000] S/m according to the rule $\sigma = 0 + i \times 500, i = 0, \ldots, 8$ and the *depth* in the range [0.05, 0.25] m, according to the rule $depth = 0.05 + i \times 0.025, i = 0, \ldots, 9$.

To achieve the optimum w^* that minimizes the discrepancy function $L()$ in (6.16), the particular characteristics of a GPR pulse has to be taken into account. Indeed, a direct matching of the time-domain scattered field leads to a high dimensionality problem (for this example a total of 1200 time steps were needed, meaning that a problem in $\mathbb{R}^{1200}$ must be solved). Also, for shallow buried cylinders and/or low dielectric contrast of soil/cylinder materials, most of the scattered signal is close to zero. Under these conditions, a direct solution of (6.16) with scattered raw data will be either inaccurate or computationally expensive.

However, with the aid of the PCA method, a remarkable reduction in the dimensionality of the problem can be achieved. Specifically, the problem reduces to a solution in $\mathbb{R}^{286}$ without any loss of information (100% of the data variance is retained), in $\mathbb{R}^{139}$ if 99.99% of the variance is kept, and in $\mathbb{R}^{51}$ if 99% of the original variance is kept. PCA is a technique originally formulated for image compression. The idea behind the method is to first orthogonalize the components of any input data (considered as uncorrelated vectors), then to order the resulting orthogonal components (called principal components) so that those with the largest variation come first in a list, and finally to eliminate those components that contribute the least to the variation in the original data set. The PCA can be thought of as a linear mapping characterized by a $p \times q$ matrix V_{pq} that transforms a given set of samples $S_p = \{x_1, \ldots, x_T | x_i \in \mathbb{R}^p\}$ into a new data set $S_q = \{y_1, \ldots, y_T | y_i \in \mathbb{R}^q\}$, where $q \leq p$. An algorithm to implement PCA is shown in Algorithm 1.

Another interpretation of the PCA is the construction of directions that maximize the variance. The transformation V_{pq} generates a projection space in which the covariance matrix is diagonal. The diagonal covariance matrix implies that the variance of a variable with itself is maximized and is minimized with respect to any other variable. Thus, the q variables with higher variance in the new space should be kept. The principal components of a set of data in $\mathbb{R}^p$ provide a sequence of best linear approximations to that data, of all ranks $q \leq p$.

Algorithm 1: PCA algorithm

function PCA(Building of V_{pq} matrix)
    for i=1 to p **do**
        subtract the mean from each data dimension,
        calculate the covariance matrix,
        calculate the eigenvectors and eigenvalues of the covariance matrix,
        choose components and form a feature vector,
        derive the new data set.
    end for
end function

The system developed in this example using a PLP network trained with the scattered wave calculated with dimensional reduction based on PCA is shown in Figure 6.12. First, given a set S of radii (ra), depths (de), ε_r, and σ, a scattered wave ($W_s \in \mathbb{R}^p$) is calculated using FDTD. Then, given $q < p$ a dimensional reduction is applied in W_s generating $W_{pca} \in \mathbb{R}^q$. Next, the W_{pca} is used in a k-fold system such that S_1 is used to estimate the expected error [Equation (6.16)], and S_{k-1} are used to adapt the PLP network parameters. In this example, a 10-fold cross-validation was employed.

To evaluate the performance of the training process, the error (loss) figure is calculated as

$$L(d_r) = \frac{|d_t - d_r|}{d_r} \tag{6.17}$$

where d is any of the unknown variables, the subscript t indicates the true value of the variable, and subscript r indicates the value reconstructed by the NN. This measures the percentage deviation of the reconstructed parameter from the actual one (desired object).

Tables 6.4 and 6.5 show the mean deviation, *mean*(L), the maximum deviation, *max*(L), and the *train* and *test* times for dimension reduction to $q = 51, 139$, and 286, when only *depth* and *radius* are considered as unknowns. This case corresponds to

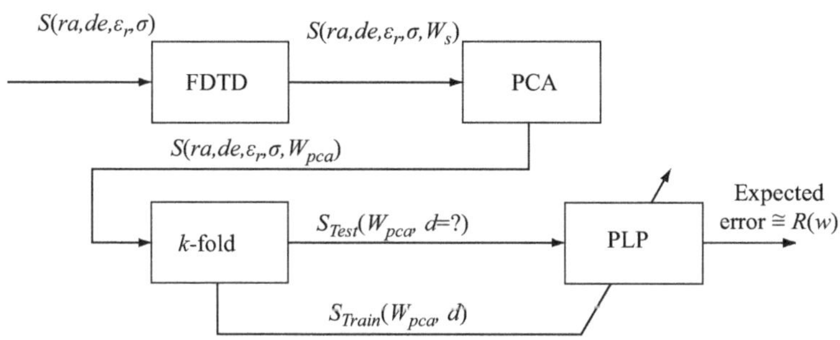

Figure 6.12 Overview of the detection system

Table 6.4 Results of the relative error for depth prediction

Configuration	*mean(L)*	max(*L*)	train(s)	test(s)
PLP(286)	0.001%	0.01%	8.44	0.016
PLP(139)	0.003%	0.15%	4.48	0.01
PLP(51)	1%	20%	6.2	0.011

L corresponds to the loss function, and train() and test() are computational times for training and validation of the ANN.

Table 6.5 Results for the relative error for radius prediction

Configuration	*mean(L)*	max(*L*)	train(s)	test(s)
PLP(286)	0.022%	0.94%	7.5	0.03
PLP(139)	1%	20%	11.1	0.018
PLP(51)	1%	22%	6.1	0.013

L corresponds to the loss function and train() and test() are computational times for training and validation of the ANN.

Table 6.6 Results for the three configurations of neural networks studied in [39]

Configuration	*Err(depth)*	*Err(radii)*
PLP(286)	5.8%	12.9%
PLP(139)	5.7%	13.4%
PLP(51)	5.7%	5.8%

the approach in [39], where three different configurations of independent and parallel networks with multiple outputs were applied to solve the same case, resulting in significant errors (Table 6.6) in comparison to the PLP network trained with the scattered waves and calculated with dimensional reduction based on PCA. Dimensional reduction was also considered in [39] but was done empirically using only three features: the peak amplitude of the reflected field, the delay of the first reflected echo calculated with respect to the time of arrival (at the receiving point) of the direct field, and a measurement of the duration of the scattered field. The average error of the best configuration presented in [39] was 1.46% for the depth reconstruction, a figure that is considerably higher than the results presented in this section.

In summary, this example outlines an approach to the characterization of inclusion in non-homogenous concrete structures using ANNs and principal component analysis. The PCA was used to preprocess the training data for the purpose of dimensional reduction. Without loss of information, it was possible to represent the initial scattered wave, $W_s \in \mathbb{R}^{1200}$, in a new set such that $W_{pca} \in \mathbb{R}^{286}$. This compressed wave was used to train a PLP to obtain a low expected error with a reasonable training time. The general system, shown in Figure 6.12, has proven to be effective for this type of problems and can be extended to other GPR problems. This example considers the non-homogeneity of the medium resulting in a more realistic model. The main advantage of ANNs compared to stochastic optimization approaches is a lower computational effort for the evaluation process [58]. As is well known, stochastic methods tend to be slow even though they are robust. The combination of both techniques should be investigated in attempts to take the advantage of the strengths of each [65].

6.3.2.1 Buried cylinder in non-homogeneous dielectric medium in the presence of noise: multi-objective artificial neural networks

One of the main difficulties in analysis of real-word GPR data is the presence of noise in the reflections from buried targets. This is due to the large variation in the inspected target signature due to environmental conditions, geometric variations, and antenna characteristics. Moreover, GPR air waves create other types of noise (such as reflections from above-surface objects) in the time window of interest. Features such as time delays and amplitudes can deteriorate in the presence of noise, yet these are fundamental to the detection and identification of buried targets. Incoherent noise can be reduced by averaging over many traces, whereas the identification and removal of coherent noise is still an important, active research topic in GPR [66].

To address some of the issues involved in detection in noisy environments, the previous example is discussed again with added noise. In this case, the problem consists of a cylindrical air inclusion buried in a non-homogenous host medium with a mean relative electrical permittivity value of 6 and standard deviation 0.25 [42,62]. The source pulse of Figure 6.6 is used as line currents in the z-direction. A sampling interval of 20 ps was chosen to meet the stability criterion of the FDTD method. A total of 1500 samples were collected for each trace corresponding to a time interval of 30 ns. The spatial steps were chosen as $\Delta x = \Delta y = 12$ mm, sufficiently small to ensure minimal numerical dispersion and resolution in modeling the inclusion. An additive white Gaussian noise is added in the simulations by modifying the scattered field.

To handle the noisy data, the ANN method is extended by preprocessing the noisy data and thus improving the SNR. This technique is based on the minimum gradient method (MGM), a multi-objective method for training NNs [67]. The resulting multi-objective NNs [68] are designed to control the complexity of the processed noisy wave, filtering undesirable frequencies dynamically by means of a network matrix Q. The process of training the NN with this matrix can be understood as a filtering process [67]. The matrix Q is a function of the network weights, $Q(w,.)$. Therefore, adapting the network weights also implies adapting the filter shape. By so doing, the risk error problem of (6.16) can be modified (in this case) to a bi-objective problem [67]:

$$\min_{w} \begin{cases} J_1(w) = R(w) \\ J_2(w) = \Omega(w) \end{cases} \tag{6.18}$$

where J_1 is related to the training error risk R, and J_2 to a complexity measure Ω, defining the filtering process required by the NN. Hence, the gradient norm can be used to account for the complexity of a given ANN in the form:

$$\min_{w} \begin{cases} J_1(w) = R(w) \\ J_2(w) = \Omega(w) = \sum_t \|\nabla_x f(x_t, w)\|^2 \end{cases} \tag{6.19}$$

where the symbol ∇_x indicates that the gradient is calculated in terms of the training parameters, even though, the control is accomplished by adjusting the weights w of the ANN. The aim of this multi-objective procedure is to reach the equilibrium between the two factors, that are, in general, conflicting. This equilibrium is responsible for the filtering capabilities of the ANN. In most cases, the issue of the dimensionality problem is more severe for this type of ANN, and techniques such as PCAs may be necessary to achieve reasonable results.

The method described here, which may be termed a PLP-MGM method (PLP combined with the MGM) is capable of removing noise from GPR data, and by so doing improves the SNR, before solving the prediction problem presented in the previous section. Figure 6.13 shows a typical result with an SNR of 3 dB. It can be seen that the proposed algorithm provides a result that is very close to the noiseless wave, therefore, improving the SNR, and, consequently, the definition of the buried object.

In the present example, the training process was conducted using an algorithm based on (6.19), with a 5-fold cross-validation. Simulations were done using 20 different runs to calculate the mean SNR as an average of the SNR for each execution. Results presented in Table 6.7 show considerable improvement in the SNR in the presence of scattered fields with 3 dB SNR.

Further validation of the procedure for real-world applications is made by considering the effect of colored Gaussian noise. This noise is a filtered version of the white noise case generated by a second-order low-pass butterworth filter with a cutoff frequency 0.4 times the bandwidth of the white noise signal. Results are shown in Table 6.8 further indicating the usefulness of this approach. Other types of noises could be considered at this point, but the PLP-MGM method seems to be robust and there is little value in pursuing these extensions.

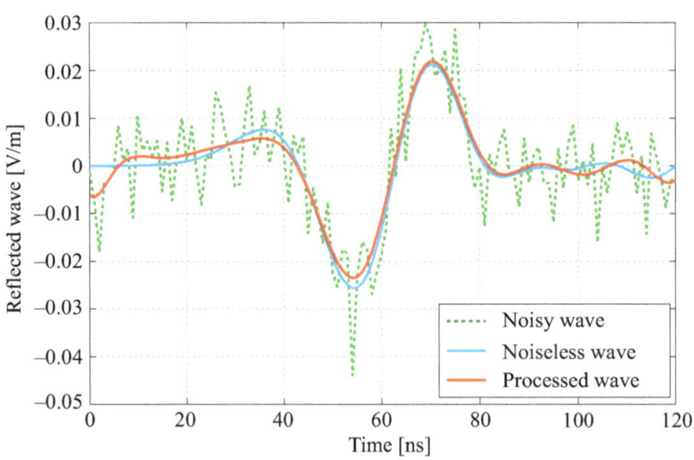

Figure 6.13 Typical reflected wave from a buried target with an SNR of 3 dB (white Gaussian noise)

Table 6.7 SNR for a wave processed (filtered) by the proposed approach corrupted by white Gaussian noise

	SNR in the filtered wave (dB)		
Noise (dB)	**Mean**	**Max**	**Min**
3	14.16	14.65	13.47
6	14.69	15.07	14.14
9	16.55	17.67	15.65
10	20.47	21.35	18.67

Table 6.8 Improvement in SNR for waves corrupted by colored Gaussian noise and filtered using the proposed PLP-MGM method

	SNR in the filtered wave (dB)		
Noise (dB)	**Mean**	**Max**	**Min**
3	12.76	13.22	11.96
6	15.30	16.21	14.17
9	20.36	20.73	19.96
10	20.58	20.93	20.09

Although this method is quite useful, the curse of dimensionality prevents its use in much more complex scenarios. The following section introduces the use of feature selection processes as alternative pattern recognition algorithms to deal with such cases.

6.3.3 Buried cylinders in concrete: feature selection

Identification of the most relevant features of datasets in time- and/or frequency-domains is a key step in the design of efficient intelligent machines for three-stage pattern recognition algorithms. The identification of the characteristics that numerically describe the samples is itself a challenging problem. Feature selection, subset selection, and variable or attribute selection [69] refer to a process commonly used in machine learning (ML), wherein a subset of the features identified in the data is selected for the application of the learning algorithm. It is different from a similar approach called feature extraction because the latter creates new features based on some combination of the original ones [70]. The best subset of features is which contains the least number of dimensions that most contribute to accuracy. Keeping a low number of dimensions that guarantees adequate convergence of the learning algorithm reduces execution time and allows the treatment of more extensive configurations. Additionally, it can provide some understanding concerning the nature of the problem, as it indicates the main physical properties needed to classify an underground target. With this approach, efficient classifiers [71] as well as regression and prediction models can be built.

There are two types of feature selection algorithms: the so-called wrapper and filter methods [72]. Wrapper methods, which are computationally intensive and tend to over-fit [73], estimate the usefulness of a subset of features by an external predictor or learning algorithm that calculates the performance of a generalization of the subset of properties. Filter or variable ranking methods compute relevance scores for each single feature and choose the most relevant based on these scores. This can be done using ad hoc evaluation functions that maximize the information for a set of features. Some of the commonly used evaluation functions are the mutual information, margin, and dependence measures.

The main drawback of these simple filtering methods is that they are not able to detect inter-feature-dependencies. One example of inter-feature-dependency in GPR is the relation between clutter and antenna interference that can occur at the same time. Neither the first nor the second dimension alone helps one to determine to which class an example belongs, only the two dimensions together contain enough information about the class membership. But inter-feature-dependencies are not the only issue. Noise contamination, redundancy or lack of samples and of features, numeric errors, incomplete registries, limited information about the problem, and inadequate preprocessing among other deficiencies can result in poor characterization of the model, more overlapping classes and instability in results of feature selection algorithms [73,74].

6.3.3.1 Simulated and experimental GPR data

In an attempt to reproduce the common configurations encountered in inspections using GPR and their system models [75], simulated data for buried inclusions in concrete are compared here to experimental GPR data. Concrete is the most common building material because it is durable, inexpensive, and readily available and can be cast into almost any shape but is replete with voids greater than 50 nm in size that are usually filled with water. Steel rebars (modeled as perfect electric conductors), and air voids (introduced during the setting process and viewed as defects) are usually found too. The material properties considered here for each medium in the simulation can be found in Table 6.9. For simulation purposes, an overall domain 0.6 m $\times$ 1 m of concrete was modeled with 2.5 mm $\times$ 2.5 mm cells, and a Gaussian pulse of central frequency set at 900 MHz and the amplitude of the line current source set at 1 A for the transmitting antenna. Transmitting and receiving antennas were located 5 cm apart at a height of 2.5 cm from the surface of the concrete over the center of the inclusion. For 2D simulations, the GprMax2D [76] was used, and both the inclusions and the solution domain are considered infinite in the transversal directions. A time window of 8 ns is used, generating 1357 samples per trace. The first 319 samples of each trace were discarded as they represent the first reflection (air–concrete interface) present in all traces. Figure 6.14 illustrates the results of simulation for a cylindrical conductor inclusion of radius 8.5 cm at the center of a concrete box and 0.28 m below its surface,

Table 6.9 Material properties used to simulate the various media

	Perfect electric conductor (PEC)	**Water**	**Free space**	**Concrete**
DC (static) relative permittivity	1	80	1	6
Relaxation time (seconds)*	0	0	0	0
DC (static) conductivity (Siemens/meter)	10^7	0.5	0	0.01
Relative permeability	1	1	1	1

*Relaxation times are zero because the media are nondispersive.

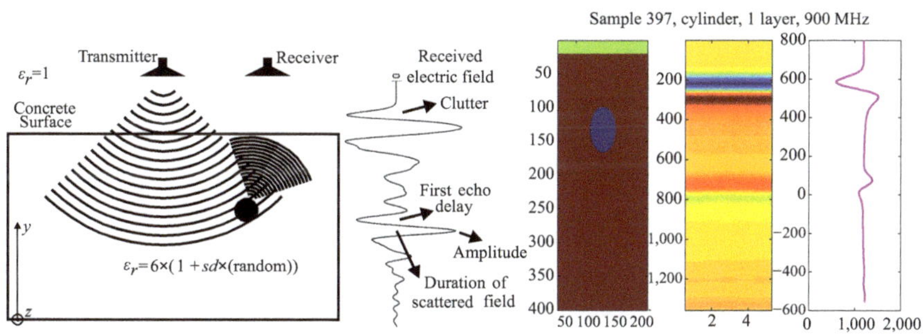

Figure 6.14 Time-domain features in a simulated reflected wave from a buried target. The figure on the right shows the B-scan radargram, whereas the center figure is an A-scan identifying some of the most relevant features of the reflected signal

showing clear identification of some relevant features (delay of the first reflection and maximum amplitude) which will be accounted for in the feature selection algorithms.

A data set of G_t pulses with 1,071 samples each was generated by concurrent execution of examples with buried water, air, and conductor cylinders at depths ranging in [0.05, 0.25] m (at increments off 0.01 m), and radii in the range [0.02, 0.1] m (at 0.005 m increments). Some noise level can be added to the resultant simulated signal for better representation of common real data from shielded antennas, as was discusses in Section 6.3.2.1. From inspection of the results, 15 features from the reflected wave were initially considered. Based on commonly used metrics for data characterization in time series and signals [77], the extracted features from the simulated signals were selected to provide information for the purpose of discrimination between classes and estimation of dimensions (some of the features are illustrated in Figure 6.15). The features selected are as follows:

In the time-domain:

1 delay of the first reflection (Delay) [62],
2 maximum amplitude (max(G_t)) [62],

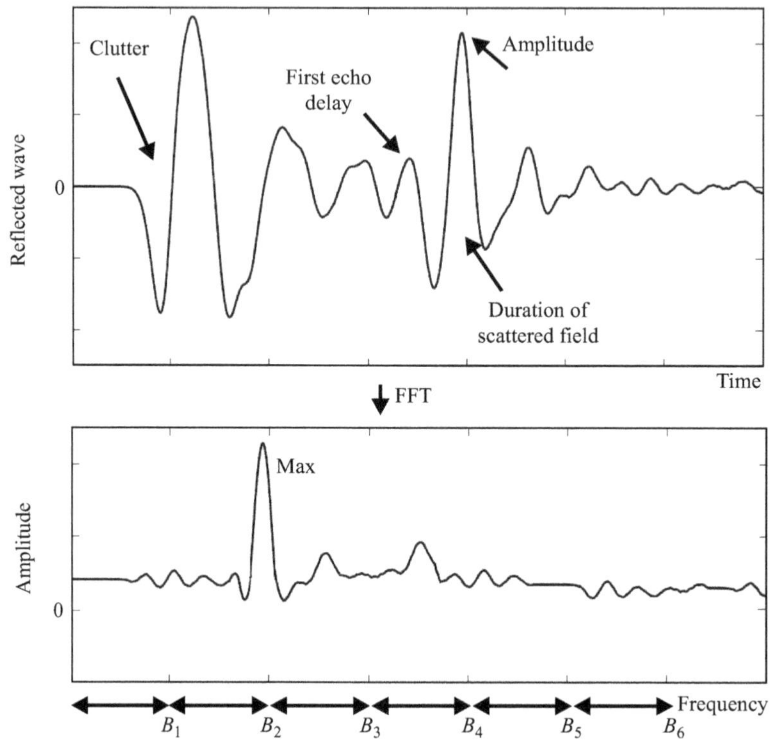

Figure 6.15 Features considered for feature selection algorithms in both the time- and frequency domain

3 reflected wave mean (mean(G_t)),
4 standard deviation of the reflected wave (std(G_t)),
5 mean of the wave derivative (mean(dG_t)),
6 standard deviation of the wave derivative (std(dG_t)),
7 signal energy (sen(G_t)).

and, by means of the Fourier Transform, in the frequency-domain:

8 maximum amplitude of the Fourier transform (mFFT),
9–14 energy in six different bands (B_1, B_2, …, B_6) of the Fourier transform, and
15 frequency of maximum amplitude of the Fourier transform (fmFFT).

Measured data were obtained from a GPR survey of concrete in a controlled, semi-anechoic chamber to avoid external noise in the experiment. The concrete samples used in the experiments described here have dimensions illustrated in Figure 6.16. The mixture has the following characteristics: (1) cement: Portland type I; (2)

water/cement ratio: 0.60; and (3) cement/sand ratio: 1:2.25. The inclusions are placed at a depth of 35 mm from the surface of the concrete slab and have diameters of 19.05 mm as illustrated in Figure 6.16. To obtain experimental results, a GPR survey was conducted on the four concrete slabs with the following characteristics: (1) without inclusion, (2) with metal inclusion, (3) with PVC inclusion, and (4) water filled inclusion.

The GPR survey equipment (Figure 6.17) consists of a 1.6 or a 2.3 GHz shielded bow-tie antenna, a control unit, XV-11 monitor with 1.2 m cable, a 4 m X3M cable, and a 11.1 V/6.6 Ah Li-Ion battery. The GPR equipment and its cable harness, along with the concrete slabs lie on an insulated support 1000 mm (± 50 mm) above the floor of the test chamber (Figure 6.18). The relative dielectric constant of the insulated support was less than 1.4. The equipment was at least 1000 mm from the chamber walls and no part of the GPR equipment was closer than 250 mm to the floor, except the battery that was located under the test bench. The cables needed by the GPR

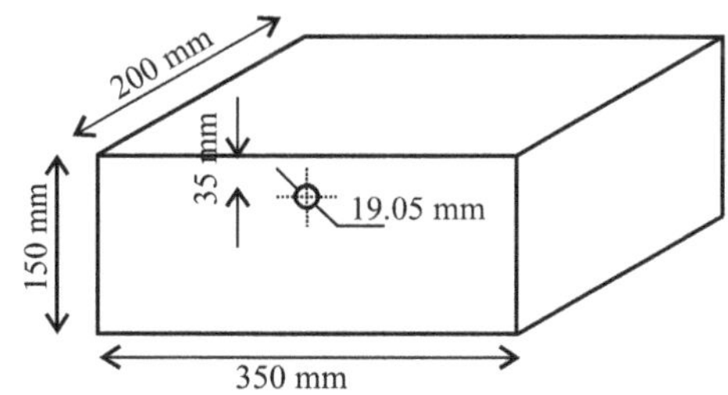

Figure 6.16 Concrete slab dimensions and location of inclusion

Figure 6.17 Experiment performed in a semi-anechoic chamber

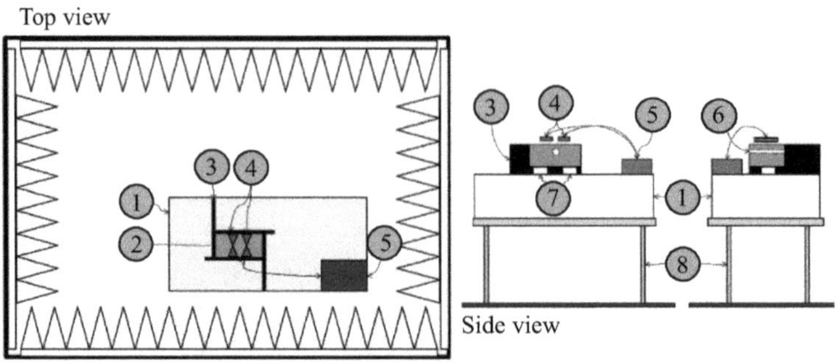

Figure 6.18 Schematic and details of the assembly placed in a semi-anechoic chamber: (1) polystyrene layer, (2) concrete block, (3) absorber plate, (4) bow-tie antennas, (5) GPR control unit, (6) inclusion, (7) polystyrene blocks, and (8) wood table

equipment to perform the tests were shielded. Some cables were excessively long and were bundled at the approximate center of the cable forming bundles 300–400 mm in length. Each concrete block was probed on its top and bottom sides. A total of 324 samples were collected with each antenna, 108 samples for each block/inclusion (air, conductor, PVC, and water). Since 12 samples for each block were taken over the inclusion, only these samples were considered with half of them taken from the top and the other half from the bottom of the blocks.

Figure 6.19 shows examples of simulated signals (with and without the noise) and of a measured signal (without processing) as a response due to a Gaussian pulse with a shielded antenna over a concrete block without inclusions. As can be noted, the signals are quite similar. Therefore, it is expected that the methodology discussed in this subsection will result in good performance even in cases of high noise contamination in GPR measurements. Under conditions of low SNR, a preprocessing step should be considered, keeping in mind that the signal attenuation due to the concrete also imposes a limit to the distinction between what should be identified and what contaminates the GPR signals.

6.3.3.2 Feature selection results

Once the collection of simulated data was complete, two feature selection algorithms from the group of variable ranking methods (Simba [78] and Relief [79]) were used to determine the essential subsets of features that improve the accuracy of the classification model and, at the same time, reduce its complexity. The Relief algorithm is based on the statistical relevance of features, whereas Simba is based on the concept of the separation margin between classes. Both methods are built on simple algorithms and produce coherent and reliable results. As a first step in both cases, it is essential to normalize the original data in each feature to avoid errors of metrics in the feature space, due to potential different scales in the magnitudes of individual features.

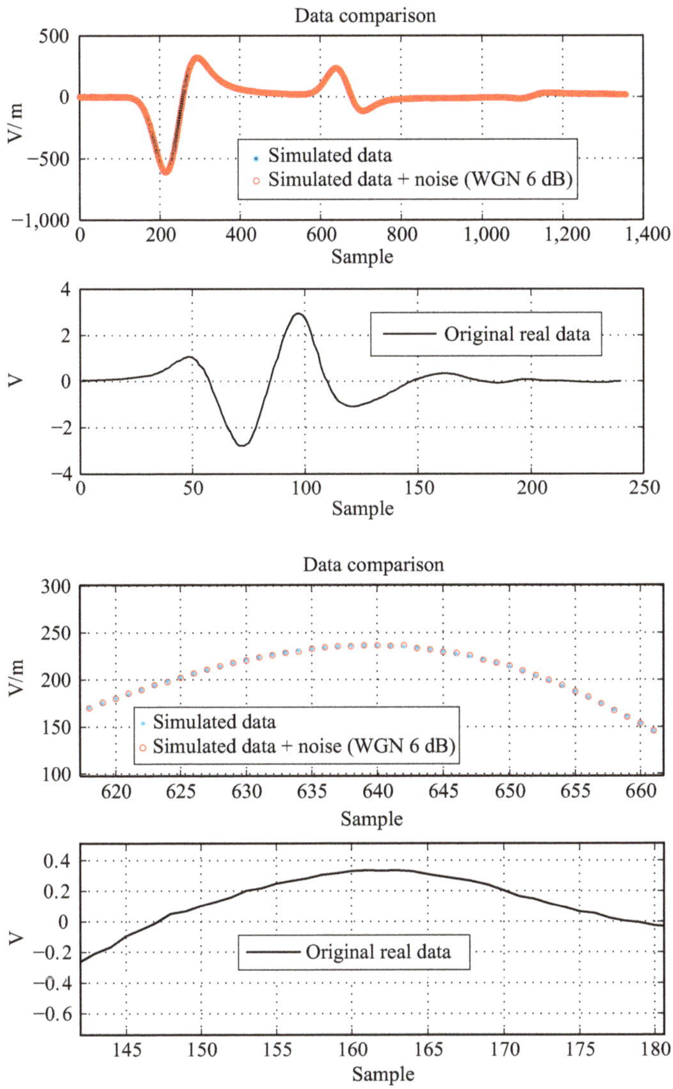

Figure 6.19 Comparison of simulated data (900 MHz) and original (unprocessed) measured data (1.6 GHz) for a Gaussian pulse for a concrete block. The two lower figures show zoomed sections of a similar section of the data for noise level comparison

The main idea of Relief, represented in the pseudocode Algorithm 2, is to compute a ranking score for every feature indicating how well this feature separates neighboring samples. The algorithm seeks the nearest neighbor from the same class (nearest hit) and from the opposite class (nearest miss) for each training sample. The Relief score for a single feature is then the difference (or ratio) between the distance to the

Algorithm 2: Relief algorithm

function RELIEF(T)
    The training set T
    A weight vector W for all features indicating their relevance.
    while until one stop criteria is achieved **do**
        $k = rand(t)$ ▷ Randomly selecting a sample x from T
        for i=1 to n **do** ▷ Finding nearest hit and miss of $x^k \in T$ in $T \setminus x^k$
            $W(i) = W(i) - dist(x_i^k, nhit(x^k))$
                $+dist(x_i^k, nmiss(x^k))$
        end for
    end while
end function

nearest miss and the distance to the nearest hit, projected on the specific feature. The algorithm generates and maintains a weight vector over all features and updates this vector for all sample points. The algorithm uses the $nhit(x^k)$, which means the nearest point in the training set excluding x^k, $T \setminus x^k$ with the same label, and the $nmiss(x)$, which implies the nearest point with the same label. These two functions provide an estimate of the margin using the 1-nearest neighbor measure where *dist* indicates the distance, usually taken as the Euclidean distance. Due to its efficiency and simple strategy for estimation of quality features, Relief has been the starting point for many other algorithms based on modifications of the original version [80–82]. As pointed in [82], the performance of Relief is affected significantly by the increase in number of features, and datasets with a significant number of irrelevant features may cause Relief to find false nearest hits and nearest misses of samples.

On the other hand, in the Simba algorithm [83], represented in pseudocode Algorithm 3, the normalized distance is calculated, as $dist_w = \sqrt{\sum_i w_i^2 x_i^2}$. Note the similarities in the pseudocodes of the two methods. However, Simba uses a stochastic gradient distance related to the number of iterations, the number of features, and the sample size, reevaluating the distances according to the weight vector W. For this reason, Relief is less efficient in eliminating redundant features [74], whereas Simba may deal with correlated features [83], provided that those features contribute to the overall performance.

In terms of computational complexity, Relief and Simba are equivalent. However, as can be seen in the output of the pseudocodes (weight vector W), each new execution produces a slightly different answer for the initial question. This can be explained by the randomly selected samples in each execution loop. Because of that the output cannot be considered unique for different subsets and even for the same group of samples. Therefore, the algorithms should be run more than once (preferably around 20 times) to obtain a consensus answer. Although the algorithm is inherently noise tolerant, the presence of noise in data may alter the results [79,81]. As a final comment, any feature selection method determines the sequence of features included in the most

Algorithm 3: Simba algorithm

function SIMBA(T)
    The training set T
    A weight vector W for all features indicating their relevance.
    while until one stop criteria is achieved **do**
        $k = rand(t)$ ▷ Randomly selecting a sample x from T
        for i=1 to n **do** ▷ Finding nearest hit and miss of $x^k \in T$ in $T \setminus x^k$ with respect to W
            $\Delta = \frac{1}{2}(dist_w(x_i^k, nhit(x)) - dist_w(x_i^k, nmiss(x^k)))$
            $W(i) = W(i) + \Delta$
        end for
        $W \leftarrow W^2/\|W^2\|_\infty$ where $(W^2)_i := (W_i)^2$
    end while
end function

discriminant subset. In particular, both Simba and Relief generated 15 subsets from 1 to 15 most discriminant GPR features, which were used to evaluate the performance of subsequent classification/regression models for the GPR dataset.

Figure 6.20 summarizes the overall flow of work for GPR predictions using feature selection. As a first step, variable ranking methods Relief and Simba provide a list of the most discriminant features based on the original set of simulated data. Each algorithm was executed 20 times and the selected features of each one were subject to a counting process that confirms the most frequent on average. At this point, the most relevant features subsets are able to feed the training-test stages of any classification/regression model one may wish to select (in this case, PLP and k-NN were applied as explained next). Of course, the choice of the number of features impacts performance and results, both in computational time and success prediction rate. For better assessment, one can alternatively apply the PCA [84] to the whole standardized feature data set. Interestingly, PCA indicated that for the original data, there is a smaller set with four principal components with each accounting for more than 2% of the total variance of the original set. In the case with noise addition, there is another group of nine principal components with each accounting for more than 2% of the total variance of the original set. These numbers support the hypothesis of using only the four most relevant features in Simba/Relief for the original data, and the nine most relevant features for noisy data.

Results of the feature selection step for the original and noisy data are outlined in Tables 6.10 and 6.11, where F_1 indicates the most relevant feature and F_4 the fourth most relevant feature. The six most often selected features were $mean(G_t)$, $mFFT$, $fmFFT$, $sen(G_t)$, $std(dG_t)$, and $delay$, which were pointed out as the most helpful in separating the clutter from the target reflections. However, in noisy situations, predictions of any model can be poorer because, for instance, estimating the amplitude is more difficult. A remarkable point is that both algorithms selected $mFFT$ and

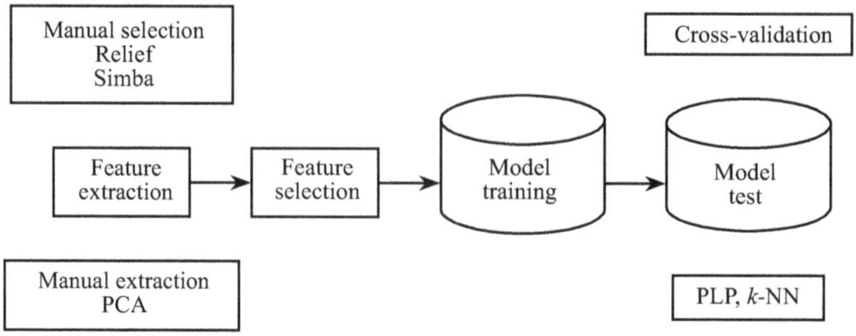

Figure 6.20 Feature extraction flow

Table 6.10 Summary of the results without noise

Inclusions	**Method**	F_1	F_2	F_3	F_4
Air/water/cond.	Simba	mean(G_t)	mFFT	Delay	fmFFT
Air/water/cond.	Relief	mFFT	mean(G_t)	sen(G_t)	std(dG_t)

Table 6.11 Summary of the results with a 6 dB SNR white Gaussian noise

Inclusions	**Method**	F_1	F_2	F_3	F_4
Air/water/cond.	Simba	mFFT	mean(G_t)	fmFFT	sen(G_t)
Air/water/cond.	Relief	mFFT	fmFFT	mean(G_t)	sen(G_t)

mean(G_t) among the four most important features for data without and with noise. The features *sen*(G_t) and *fmFFT* are present in three of the four sets of the most important features. Almost all of them are related to the signal as a whole in both sets of data and not to time segments of the signals. The rest of the features were assigned different order of importance by each algorithm and for different dataset noise conditions as expected. During the 20-execution feature selection process, it became clear that Relief results are more stable than Simba for the original dataset. After noise addition, the results from both algorithms were affected significantly, but Relief still showed more homogeneous results.

Once the most relevant features were identified, the next step is to characterize the inclusions in concrete in terms of three different characteristics: (1) the material (air/water/conductor), (2) depth, and (3) radius. The k-NN classifier model was used to characterize the material, whereas regression models were used to characterize depth and radius and to assess the numerical values for each. To estimate the performance of the feature selection stage, results were obtained for all models with different runs, increasing the number of features under consideration from 1 to 15.

The performance of the k-NN classification model was evaluated according to an average accuracy from a multi-class confusion matrix, whereas for the regression model the root-mean-square error (RMSE) was used to allow comparisons.

Results for the material classification of the inclusions are shown in Tables 6.12 and 6.13, which show the results of the k-NN model for signals without and with noise using selected features from Relief and Simba. Column 1 shows the classification type, followed by the number of features considered in the experiment (column 2) and the average values of accuracy for the classification models (columns 3 and 4). It is very interesting to remark that accuracy for the original data with both Relief and Simba algorithms reaches excellent performance when the number of features reaches 4 (the same number predicted by the PCA algorithm). Including more features increases the computational time without significant improvement in results. However, for classifications under noisy conditions (which exhibit lower performance overall), increasing the number of features results in lower accuracies (Figure 6.21). This result may indicate that some of the features are affected by noise and could be discarded without loss in performance. Finally, even though for the original data Relief models are not better than Simba models, in most cases they are at least equal in the case of signal with noise, which can possibly be due to the improved robustness of Relief mentioned earlier.

Regarding the specific details of the classifier employed, a k-NN [85] model using Euclidean distance to the $k = 3$ nearest labeled points was used to classify 21 or 22 testing points of 1050 or 1049 training samples using a $k = 50$-fold cross-validation

Table 6.12 Classification of results for features selected by Relief

Inclusions	**Number of feat.**	**Accuracy (orig.)**	**Accuracy (noise)**
Air/water/cond.	1	$67 \pm 9\%$	$61 \pm 10\%$
Air/water/cond.	4	$96 \pm 4\%$	$83 \pm 9\%$
Air/water/cond.	7	$98 \pm 3\%$	$82 \pm 8\%$
Air/water/cond.	10	$100 \pm 1\%$	$78 \pm 9\%$
Air/Water/Cond.	13	$99 \pm 2\%$	$72 \pm 11\%$
Air/Water/Cond.	15	$99 \pm 2\%$	$72 \pm 10\%$

Table 6.13 Classification of results for features selected by Simba

Inclusions	**Number of Feat.**	**Accuracy (orig.)**	**Accuracy (noise)**
Air/water/cond.	1	$80 \pm 7\%$	$61 \pm 11\%$
Air/water/cond.	4	$99 \pm 2\%$	$82 \pm 8\%$
Air/water/cond.	7	$100 \pm 1\%$	$81 \pm 8\%$
Air/water/cond.	10	$100 \pm 1\%$	$75 \pm 9\%$
Air/water/cond.	13	$100 \pm 1\%$	$73 \pm 9\%$
Air/water/cond.	15	$99 \pm 2\%$	$72 \pm 8\%$

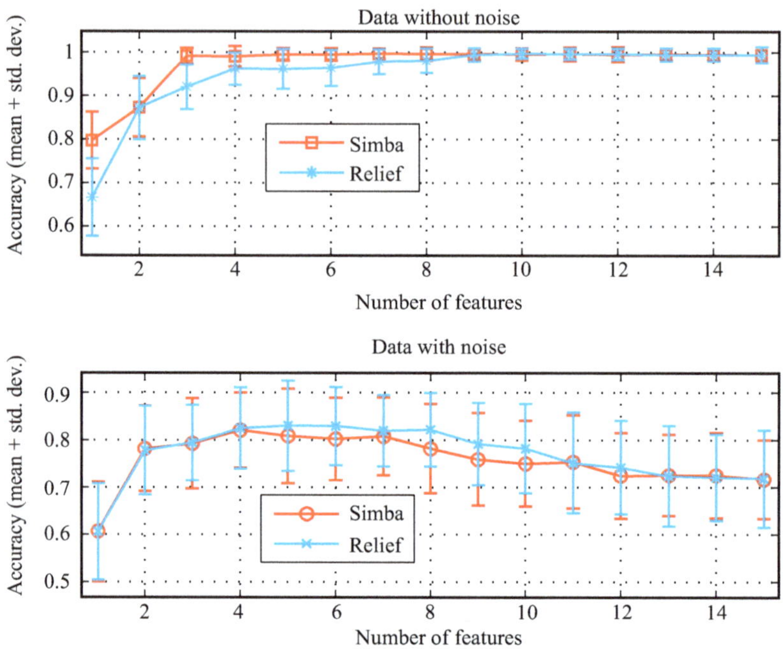

Figure 6.21 Comparative performances of k-NN models with different number of features selected by Simba and Relief for data without noise and with the addition of noise

with balanced classes (water/air/conductor). The number of neighbors for k-NN was chosen to obtain the smallest models with acceptable results in the verification stage.

The results with *PLP* regression models for depth are shown in Tables 6.14 and 6.15, whereas Tables 6.16 and 6.17 show the results with *PLP* regression models for radius. Discarding the "Delay" feature, classified as the least relevant for noisy signals, the respective percentage errors using the RMSE range from −1.6% to 1.1% for radius and from −38% to −0.1% for depth. Although the increase in number of features in models for depth generates much better results for the original data, no significant performance enhancement occurs beyond models with five features for data with noise. The order in which features were added in both algorithms had similar mean RMSE for noisy signals. The improvement for Simba models was faster for the original data, but Relief models reached the lowest mean RMSE values using fewer features. Also, the mean RMSE values for radius models were significantly lower than those for depth models.

Both *PLP* regression models [67] were trained with the same number of samples for depth and radius outputs. Each desired output of these two parameters was evaluated with separate models with two linear neurons and two neurons with sigmoidal

Table 6.14 Regression results for output 1 (depth—(RMSE) in meters) for features selected by Relief

Inclusions	Number of Feat.	RMSE (orig.)	RMSE (noise)
Air/water/cond.	1	0.052 ± 0.004	0.052 ± 0.004
Air/water/cond.	4	0.036 ± 0.008	0.040 ± 0.007
Air/water/cond.	7	0.03 ± 0.01	0.040 ± 0.008
Air/water/cond.	10	0.019 ± 0.004	0.040 ± 0.009
Air/water/cond.	13	0.02 ± 0.02	0.039 ± 0.006
Air/water/cond.	15	0.020 ± 0.006	1 ± 8

Table 6.15 Regression results of output 1 (depth—(RMSE) in meters) for features selected by Simba

Inclusions	Number of Feat.	RMSE (orig.)	RMSE (noise)
Air/water/cond.	1	0.06 ± 0.01	0.052 ± 0.004
Air/water/cond.	4	0.022 ± 0.003	0.04 ± 0.02
Air/water/cond.	7	0.022 ± 0.008	0.039 ± 0.006
Air/water/cond.	10	0.02 ± 0.02	0.040 ± 0.005
Air/water/cond.	13	0.019 ± 0.006	0.039 ± 0.008
Air/water/cond.	15	0.019 ± 0.005	4 ± 32

Table 6.16 Regression results of output 2 (radius—(RMSE) in meters) for features selected by Relief

Inclusions	Number of Feat.	RMSE (orig.)	RMSE (noise)
Air/water/cond.	1	0.023 ± 0.002	0.023 ± 0.002
Air/water/cond.	4	0.018 ± 0.005	0.021 ± 0.002
Air/water/cond.	7	0.017 ± 0.004	0.020 ± 0.002
Air/water/cond.	10	0.016 ± 0.003	0.020 ± 0.003
Air/water/cond.	13	0.016 ± 0.004	0.021 ± 0.003
Air/water/cond.	15	0.016 ± 0.006	1 ± 5

activation function in the hidden layer. The number of training and testing samples was the same. The maximum permitted number of epochs was 10 and the mean squared error (MSE) was 0.001 for the convergence test of the Levenberg–Marquardt [86,87] hybrid training method (for nonlinear parameters—the Levenberg–Marquardt algorithm is a method of solution for nonlinear problems in least squares) and least square estimate (for linear parameters). The number of hidden neurons for the PLP was chosen to obtain the smallest models with acceptable results, because computational efficiency was not considered a goal of this test.

Table 6.17 Regression results of output 2 (radius—(RMSE) in meters) for features selected by Simba

Inclusions	Number of Feat.	RMSE (orig.)	RMSE (noise)
Air/water/cond.	1	0.024 ± 0.003	0.023 ± 0.002
Air/water/cond.	4	0.022 ± 0.003	0.022 ± 0.006
Air/water/cond.	7	0.017 ± 0.004	0.021 ± 0.003
Air/water/cond.	10	0.017 ± 0.004	0.021 ± 0.003
Air/water/cond.	13	0.016 ± 0.003	0.020 ± 0.003
Air/water/cond.	15	0.015 ± 0.002	0.1 ± 0.4

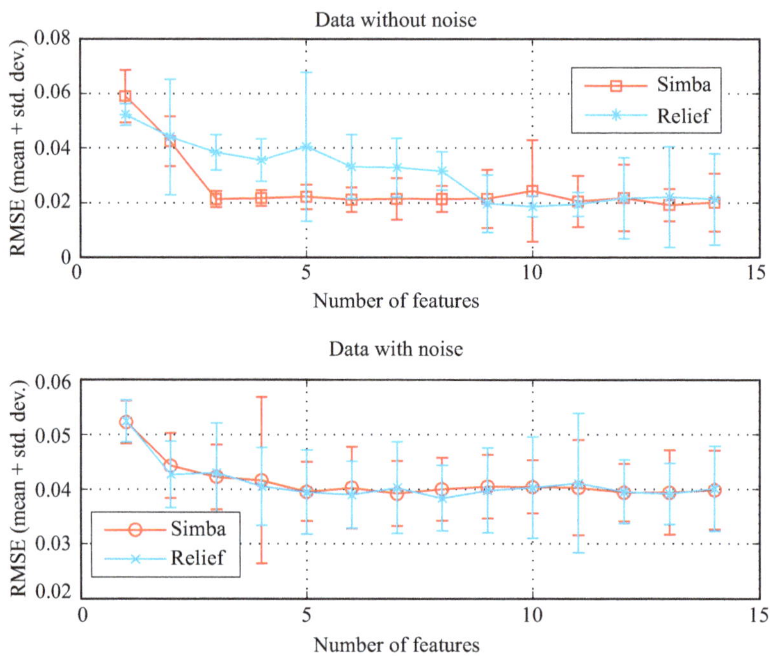

Figure 6.22 RMSE results of PLP regression model for depth

Figures 6.22 and 6.23 show the evolution of the RMSE as the number of features increases (the last feature, "Delay" was removed for visualization purposes). For noisy data, Relief models provide better results, although the evolution of the mean RMSE values for both algorithms and both conditions of noise are quite similar. As expected, the accuracy of the models for data with noise is worse than for the original data. This is partially due to the difficulty in extracting the values for each feature from raw data with noise. This fact suggests to recommend preprocessing of the data before extracting the features, through ad hoc numerical filters designed to obtain

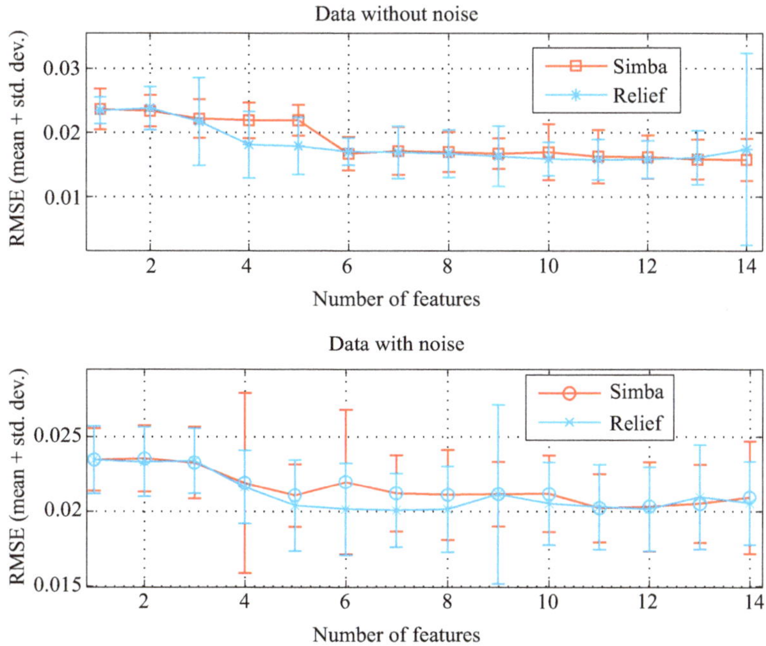

Figure 6.23 RMSE results of PLP regression model for radius

amplitude and phase responses which could remove the noise without distorting the original data.

In summary, the methodology presented here shows that once an inclusion is detected in an experimental radargram, PRAs based on feature selection considering only simulated datasets and some extracted metrics are useful in discriminating material composition, depth, and dimensions of the inclusions. Even though this model has limitations (for instance, in cases of existence of many inclusions in the scenario), the potential of pattern recognition algorithms for GPR is fully demonstrated.

6.4 Summary

This chapter has briefly discussed the use of pattern recognition techniques for GPR applications. Pattern recognition methods attempt to determine a useful finite number of parameters that characterize targets in a given medium by means of identification of their electrical and geometrical properties. In general, this is a challenging problem because the available data are often insufficient for the prediction. In some cases, this lack of data can be addressed through the insertion of additional information on the GPR scenario, such as symmetries or other simplifying assumptions. In other cases, a large number of data are available but, if noise and other external sources are present in the data, the information contains uncertainties that can cause the prediction system to be unstable or produce inaccurate solutions.

GPR has attracted considerable interest from the NDE community. As more applications for GPR as an NDE tool become practical, the complexity of the post-processing algorithms continues to grow. The use of ANN and ML techniques is a significant milestone because the new approach tackles the important problem of interpretation of GPR data with minimum or no human assistance. An important challenge is to design algorithms that can resolve uncertainty related to false-positives and false-negatives. Another common challenge is the improvement of object localization, handling of multiple objects, and the ever increasing need for higher resolution.

References

[1] Andre F, Jonard M, and Lambot S. Non-invasive forest litter characterization using full-wave inversion of microwave radar data. IEEE Transactions on Geoscience and Remote Sensing, 53 (2);2015:828–840.
[2] Soldovieri F, Gennarelli G, Catapano I, Liao D, and Dogaru T. A comparison of two data processing strategies. IEEE Journal of Selected Topics in Applied Earth Observations and Remote Sensing, 10 (2);2017:562–571.
[3] Hagedoorn JG. A process of seismic reflection interpretation. Geophysical Prospecting, 2 (2);1954:85–127.
[4] Gazdag J and Sguazzero P. Migration of seismic data. Proceedings of the IEEE, 72;1984:1302–1315.
[5] Berkhout AJ. Wave field extrapolation techniques in seismic migration – A tutorial. Geophysics, 46;1981:1638–1656.
[6] Berkhout AJ. Seismic Migration-Imaging of Acoustic Energy by Wave Field Extrapolation – A. Theorectical Aspects. Elsevier, New York; 1982.
[7] Leuschen CJ and Plumb RG. A matched-filter-based reverse-time migration algorithm for ground-penetrating radar data. IEEE Trans Geoscience and Remote Sensing, 39;2001:929–936.
[8] Zhu J, Xue Y, Zhang N, Li Z, Tao Y, and Qiu D. A noise reduction method for ground penetrating radar signal based on wavelet transform and application in tunnel lining. IOP Conference Series: Earth and Environmental Science, 61;2017:012088.
[9] Williams A. Convolutional Neural Networks in Python: Introduction to Convolutional Neural Networks. CreateSpace Independent Publishing Platform, Jakarta; 2017. doi: 10.1109/ICRAMET.2017.8253139.
[10] Al-Saffar AAM, Tao H, and Talab MA. Review of deep convolution neural network in image classification. In: 2017 International Conference on Radar, Antenna, Microwave, Electronics, and Telecommunications (ICRAMET); 2017:26–31.
[11] Reichman D, Collins LM, and Malof JM. Some good practices for applying convolutional neural networks to buried threat detection in ground penetrating

radar. In: 2017 9th International Workshop on Advanced Ground Penetrating Radar (IWAGPR); 2017:1–5.

[12] Noreen T and Khan US. Using pattern recognition with HOG to automatically detect reflection hyperbolas in ground penetrating radar data. In: 2017 International Conference on Electrical and Computing Technologies and Applications (ICECTA); 2017:1–6.

[13] Khalaf MW, El-Hefnawi FM, and Harb HM. Identification and classification of reflected landmine signals based on complex resonance frequency in dispersive media. In: 2017 Intl Conf on Advanced Control Circuits Systems (ACCS) Systems 2017 Intl Conf on New Paradigms in Electronics Information Technology (PEIT); 2017:223–231.

[14] Ajithkumar N, Aswathi P, and Bhavani RR. Identification of an effective learning approach to landmine detection. In: 2017 1st International Conference on Electronics, Materials Engineering and Nano-Technology (IEMENTech); 2017:1–5.

[15] Shawe-Taylor J and Cristianini N. Kernel Methods for Pattern Analysis. Kernel Methods for Pattern Analysis. Cambridge University Press, Cambridge; 2004.

[16] Sun Y and Li J . Time-frequency analysis for plastic landmine detection via forward-looking ground penetrating radar. IEE Proceedings – Radar Sonar and Navigation, 150;2003.

[17] Gader PD, Mystkowski M, and Zhao Y. Landmine detection with ground penetrating radar using hidden Markov models. IEEE Transactions on Geoscience and Remote Sensing, 39 (6);2001:1231–1244.

[18] Manandhar A, Torrione PA, Collins LM, and Morton D. Multiple-instance hidden Markov model for GPR-based landmine detection. IEEE Transactions on Geoscience and Remote Sensing, 53 (4);2015:1737–1745.

[19] Gader PD and Mystkowski M. Applications of hidden Markov models to detecting land mines with ground-penetrating radar. Proceedings of Aerospace '99, Volume 3710, Detection and Remediation Technologies for Mines and Minelike Targets IV, Orlando, FL; 1999. https://doi.org/10.1117/12.356989.

[20] Missaoui O, Frigui H, and Gader P. Land-mine detection with ground-penetrating radar using multistream discrete hidden Markov models. IEEE Transactions on Geoscience and Remote Sensing, 49 (6);2011:2080–2099.

[21] Duval F. Artificial Neural Networks: Concepts, Tools and Techniques Explained for Absolute Beginners. Data Sciences. CreateSpace Independent Publishing Platform; 2018.

[22] Rashid T. Make Your Own Neural Network: A Gentle Journey Through the Mathematics of Neural Networks, and Making Your Own Using the Python Computer Language. CreateSpace Independent Publishing Platform, Chichester; 2016.

[23] McCann MT, Jin KH, and Unser M. Convolutional neural networks for inverse problems in imaging: A review. IEEE Signal Processing Magazine, 34 (6);2017:85–95.

[24] Torres-Huitzil C, and Girau B. Fault and error tolerance in neural networks: A review. IEEE Access, 5;2017:17322–17341.

[25] Huang G, Huang GB, Song S, and You K. Trends in extreme learning machines: A review. Neural Networks, 61;2015:32–48.

[26] Bayoudh K. From Machine Learning to Deep Learning, Goodreads; 2017.

[27] Nickel M, Murphy K, Tresp V, and Gabrilovich. A review of relational machine learning for knowledge graphs. Proceedings of the IEEE, 104 (1);2016:11–33.

[28] Choudhary R and Gianey HK. Comprehensive review on supervised machine learning algorithms. In: 2017 International Conference on Machine Learning and Data Science (MLDS); 2017:37–43.

[29] Al-Jarrah OY, Yoo PD, Muhaidat S, Karagiannidis GK, and Taha K. Efficient machine learning for big data: A review. Big Data Research, 2 (3);2015: 87–93. Big Data, Analytics, and High-Performance Computing. Available from: http://www.sciencedirect.com/science/article/pii/S2214579615000271.

[30] Costamagna E, Gamba P, and Losani S. A neural network approach to the interpretation of ground penetrating radar data. In: Geoscience and Remote Sensing Symposium Proceedings, 1998. IGARSS '98. 1998 IEEE International. vol. 1; 1998:412–414.

[31] Bralich J, Reichman D, Collins LM, and Malof JM. Improving convolutional neural networks for buried target detection in ground penetrating radar using transfer learning via pretraining. Proceedings of SPIE Defense and Security, 2017, Volume 10182, Detection and Sensing of Mines, Explosive Objects, and Obscured Targets XXII; 101820X, Anaheim CA; 2017. https://doi.org/10.1117/12.2263112.

[32] Jo H, Na Y, and Song J. Data augmentation using synthesized images for object detection. In: 2017 17th International Conference on Control, Automation and Systems (ICCAS); 2017:1035–1038.

[33] Ding J, Li X, and Gudivada VN. Augmentation and evaluation of training data for deep learning. In: 2017 IEEE International Conference on Big Data (Big Data); 2017:2603–2611.

[34] Gamba P and Lossani S. Neural detection of pipe signatures in ground penetrating radar images. IEEE Transactions on Geoscience and Remote Sensing, 38 (2);2000:790–797.

[35] Gader PD, Keller JM, and Nelson BN. Recognition technology for the detection of buried land mines. IEEE Transactions on Fuzzy Systems, 9 (1);2001:31–43.

[36] Shaw MR, Millard SG, Molyneaux TCK, Taylor MJ, and Bungey JH. Location of steel reinforcement in concrete using ground penetrating radar and neural networks. NDT E International, 38 (3);2005:203–212. Structural Faults and Repair.

[37] Moysey S, Caers J, Knight R, and Allen-King RM. Stochastic estimation of facies using ground penetrating radar data. Stochastic Environmental Research and Risk Assessment (SERRA), 17 (5);2003:306–318. Available from: https://doi.org/10.1007\%2Fs00477-003-0152-6.

[38] Gilmore C, Peters S, LoVetri J, and McNeil D. GPR target detection using a neural network classifier of image moments as invariant features. In: 2004 10th International Symposium on Antenna Technology and Applied Electromagnetics and URSI Conference; 2004:1–4.

[39] Caorsi S and Cevini G. An electromagnetic approach based on neural networks for the GPR investigation of buried cylinders. IEEE Geoscience and Remote Sensing Letters, 2 (2)2005:3–7.

[40] Yang, CC and Bose NK. Landmine detection and classification with complex-valued hybrid neural network using scattering parameters dataset. IEEE Transactions on Neural Networks, 16 (3); 2005:743–753.

[41] Lee W, Gader PD, and Wilson JN. Optimizing the area under a receiver operating characteristic curve with application to landmine detection. IEEE Transactions on Geoscience and Remote Sensing, 45 (2);2007:389–397.

[42] Travassos XL, Vieira DAG, Ida N, Vollaire C, and Nicolas A. Characterization of inclusions in a nonhomogeneous GPR problem by artificial neural networks. IEEE Transactions on Magnetics, 44 (6);2008:1630–1633.

[43] Cao Y, Guzina BB, and Labuz JF. Mobile Systems IV. Department of Civil Engineering, University of Minnesota; 2008. MN/RC 2008-10. Available from: https://www.lrrb.org/pdf/200810.pdf.

[44] Travassos XL, Vieira DAG, Palade V, and Nicolas A. Noise reduction in a non-homogenous ground penetrating radar problem by multiobjective neural networks. IEEE Transactions on Magnetics, 45 (3);2009:1454–1457.

[45] Singh NP and Nene MJ. Buried object detection and analysis of GPR images: Using neural network and curve fitting. In: 2013 Annual International Conference on Emerging Research Areas and 2013 International Conference on Microelectronics, Communications and Renewable Energy; 2013:1–6.

[46] Qiao X, Yang F, and Xu X. The prediction method of soil moisture content based on multiple regression and RBF neural network. In: Proceedings of the 15th International Conference on Ground Penetrating Radar; 2014:140–143.

[47] Nieto X, Solla M, Gómez-Perez P, and Lorenzo H. GPR signal characterization for automated landmine and UXO detection based on machine learning techniques. Remote Sensing, 10 (6);2014.

[48] Szymczyk P and Szymczyk M. Classification of geological structure using ground penetrating radar and Laplace transform artificial neural networks. Neurocomputing, 148; 2015:354–362.

[49] Tomecka-Suchoń S, Szymczyk P, and Szymczyk M. Neural networks as a tool for georadar data processing. International Journal of Applied Mathematics, 25;2015:955–960.

[50] Sakaguchi RT, Morton KD, Collins LM, and Torrione PA. Recognizing subsurface target responses in ground penetrating radar data using convolutional neural networks. Proceedings of SPIE Defense and Security, 2015, Volume 9454, Detection and Sensing of Mines, Explosive Objects, and Obscured Targets XX; 94541A, Baltimore MD; 2015. https://doi.org/10.1117/12.2177747.

[51] Zhao S and Al-Qadi IL. Development of regularization methods on simulated ground-penetrating radar signals to predict thin asphalt overlay thickness. Signal Processing, 132;2017:261–271.

[52] Sakaguchi R, Crosskey M, Chen D, Walenz B, and Morton Jr. K. Convolutional neural network based sensor fusion for forward looking ground penetrating radar. Proceedings of SPIE Defense and Security, 2016, Volume 9823,

Detection and Sensing of Mines, Explosive Objects, and Obscured Targets XXI; 98231J, Baltimore, MD; 2016. https://doi.org/10.1117/12.2224125.

[53] Dou Q, Wei L, Magee DR, and Cohn AG. Real-time hyperbola recognition and fitting in GPR data. IEEE Transactions on Geoscience and Remote Sensing, 55 (1); 2017:51–62.

[54] Xu M, Fan C, Paden JD, Fox GC, and Crandall DJ. Multi-task spatiotemporal neural networks for structured surface reconstruction. In: 2018 IEEE Winter Conference on Applications of Computer Vision (WACV); 2018:1273–1282.

[55] Jia H, Takenaka T, and Tanaka T. Time-domain inverse scattering method for cross-borehole radar imaging. IEEE Transactions on Geoscience and Remote Sensing, 40 (7);2002:1640–1647.

[56] Deming RW and Devaney AJ. Diffraction tomography for multi-monostatic ground penetrating radar imaging. Inverse Problems, 13 (1);1997:29–45.

[57] Daniels DJ. Ground Penetrating 2nd ed. Institution of Engineering and Technology, London; 2004.

[58] Travassos XL, Vieira DAG, Ida N, Vollaire C, and Nicolas A. Inverse algorithms for the GPR assessment of concrete structures. IEEE Transactions on Magnetics, 44 (6); 2008:994–997.

[59] Eberhart RC, Simpson PK, and Dobbins RW. Computational Intelligence PC Tools. Academic Press Professional, Boston, MA; 1996. Chapter 5.

[60] GSSI Handbook For Radar Testing of Concrete. Published by Geophysical Survey Systems, Inc., North Salem, NH; 2001. p. 13.

[61] Guidance on Radar Testing of Concrete Structures. The Concrete Society; 1997. http://www.concrete.org.uk.

[62] Caorsi S and Cevini G. Neural networks trained by scattered electromagnetic data for GPR applications. In: 2nd International Workshop on Advanced GPR. The Netherlands; 2003:14–16.

[63] Caminhas WM, Vieira DAG, and Vasconcelos JA. Parallel layer perceptron. Neurocomputing, 55;2003:771–778.

[64] Vapnik VN. The Nature of Statistical Learning Theory (Statistics for Engineering and Information Science). 2nd ed. Springer, New York, NY; 2001.

[65] Vieira DG, Vieira DAG, Caminhas WM, and Vasconcelos JA. A hybrid approach combining genetic algorithm and sensitivity information extracted from a parallel layer perceptron. IEEE Transactions on Magnetics, 41 (5); 2005:1740–1743.

[66] Nuzzo L and Quarta T. Improvement in GPR coherent noise attenuation using tau and wavelet transforms. Geophysics, 69;2004.

[67] Vieira DAG, Takahashi RHC, PaladeV, Vasconcelos JA, and Caminhas WM. The Q-norm complexity measure and the minimum gradient method: A novel approach to the machine learning structural risk minimization problem. IEEE Transactions on Neural Networks, 19 (8);2008:1415–1430.

[68] Vieira D, de Vasconcelos J, and Caminhas WM. Controlling the parallel layer perceptron complexity using a multiobjective learning algorithm. Neural Computing and Applications, 16;2007:317–325.

[69] Liu H and Yu L. Toward integrating feature selection algorithms for classification and clustering. IEEE Transactions on Knowledge and Data Engineering, 17 (4);2005:491–502.

[70] Dash M and Liu H. Dimensionality Reduction. Wiley Encyclopedia of Computer Science and Engineering, New York, NY; 2007. doi:10.1002/9780470050118.ecse112.

[71] Liu H. Evolving feature selection. IEEE Intelligent Systems and Their Applications, 20 (6);2005:64.

[72] Jain A and Zongker D. Feature selection: evaluation, application, and small sample performance. IEEE Transactions on Pattern Analysis and Machine Intelligence, 19 (2);1997:153–158.

[73] Křížek P, Kittler J, and Hlaváč V. Improving Stability of Feature Selection Methods. In: Kropatsch WG, Kampel M, and Hanbury A, editors. Computer Analysis of Images and Patterns. Berlin, Heidelberg: Springer Berlin Heidelberg; 2007. p. 929–936.

[74] Kalousis A, Prados J, and Hilario M. Stability of feature selection algorithms: A study on high-dimensional spaces. Knowledge and Information Systems, 12 (1);2007:95–116.

[75] Warren C and Giannopoulos A. Creating finite-difference time-domain models of commercial ground-penetrating radar antennas using Taguchi's optimization method. Geophysics, 76;2011:G37–G47.

[76] Giannopoulos A. Modelling ground penetrating radar by GprMax. Construction and Building Materials, 19 (10);2005:755–762. Non Destructive Testing: Selected papers from Structural Faults and Repair 2003.

[77] Oppenheim AV, Willsky AS, and Nawab SH. Signals & Systems, 2nd ed. Prentice-Hall, Inc., Upper Saddle River, NJ, USA; 1996.

[78] Guyon I, Gunn S, Nikravesh M, and Zadeh LA, editors. Feature Extraction, Foundations and Applications. Series Studies in Fuzziness and Soft Computing. Springer-Verlag, Berlin; 2006.

[79] Kira K and Rendell LA. A practical approach to feature selection. In: Proceedings of the Ninth International Workshop on Machine Learning. ML92. San Francisco, CA, USA: Morgan Kaufmann Publishers Inc.; 1992:249-256. Available from: http://dl.acm.org/citation.cfm-id=141975.142034.

[80] Sun Y and Li J. Iterative RELIEF for feature weighting. In: Proceedings of the 23rd International Conference on Machine Learning. ICML '06. New York, NY, USA: ACM; 2006:913–920. Available from: http://doi.acm.org/10.1145/1143844.1143959.

[81] Kononenko I. Estimating attributes: Analysis and extensions of RELIEF. In: Bergadano F and De Raedt L, editors. Machine Learning: ECML-94. Berlin, Heidelberg: Springer Berlin Heidelberg; 1994. p. 171–182.

[82] Sun Y, Todorovic S, and Goodison S. Local-learning-based feature selection for high-dimensional data analysis. IEEE Transactions on Pattern Analysis and Machine Intelligence, 32 (9);2010:1610–1626.

[83] Gilad-Bachrach R, Navot A, and Tishby N. Margin based feature selection – Theory and algorithms. In: Proc of the 21th Int Conf on Machine Learning. vol. 9;2004.
[84] Martinez AM and Kak AC. PCA versus LDA. IEEE Transactions on Pattern Analysis and Machine Intelligence, 23 (2);2001:228–233.
[85] Mitchell TM. Machine Learning, 1st ed. McGraw-Hill, Inc., New York, NY, USA; 1997.
[86] Levenberg K. A method for the solution of certain non-linear problems in least squares. Quarterly of Applied Mathematics, 2 (2);1944:164–168.
[87] Marquardt D. An algorithm for least-squares estimation of nonlinear parameters. SIAM Journal on Applied Mathematics, 11 (2);1963:431–441.

Index

www.ingramcontent.com/pod-product-compliance
Lightning Source LLC
LaVergne TN
LVHW051050210926
848486LV00001B/58

* 9 7 8 1 7 8 5 6 1 4 9 3 4 *